Geography Skills for NCEA Level One

2nd Edition

Justin Peat

Australia • Brazil • Japan • Korea • Mexico • Singapore • Spain • United Kingdom • United States

Geography Skills for NCEA Level One
2nd Edition
Justin Peat

Cover design: Smartwork Creative Ltd
Text designer: Smartwork Creative Ltd
Production controller: Siew Han Ong

Any URLs contained in this publication were checked for currency during the production process. Note, however, that the publisher cannot vouch for the ongoing currency of URLs.

Acknowledgements
The authors and publisher wish to thank the following people and organisations for permission to use the resources in this textbook. Every effort has been made to trace and acknowledge all copyright owners of material used in this book. However, if any infringement has occurred the publishers tender their apologies and invite the copyright holders to contact them.
Getty images: cover image

Shutterstock: pages 11, 13, 28 (bottom), 71 (bottom), 72, 73, 74, 75, 76, 88 (bottom), 96, 98, 99, 100, 102, 110 and 115.

Pages 9, 10, 12, 17, 19, 25, 27, 31, 34, 36, 104, 100, 104 and 107 courtesy of Land Information New Zealand; page 15 courtesy of Air New Zealand; page 18 illustration courtesy of Cheryl Rowe; pages 29 and 70 illustrations courtesy of Brenda Cantell; page 41 map (bottom) courtesy of NZSki Limited; page 71 satellite image courtesy of Geo Eye; pages 77, 79, 80, 114 and 119 cartoons courtesy of Alexander Turnbull Library, Wellington, NZ; page 106 image courtesy of GNS Science; pages 116 and 117 maps courtesy of Auckland Council.

For product information and technology assistance,
in Australia call **1300 790 853**;
in New Zealand call **0800 449 725**

For permission to use material from this text or product, please email **aust.permissions@cengage.com**

National Library of New Zealand Cataloguing-in-Publication Data
A catalogue record for this book is available from the National Library of New Zealand.

9780170368155

Cengage Learning Australia
Level 7, 80 Dorcas Street
South Melbourne, Victoria Australia 3205

Cengage Learning New Zealand
Unit 4B Rosedale Office Park
331 Rosedale Road, Albany, North Shore 0632, NZ

For learning solutions, visit **cengage.co.nz**

Printed in Singapore by C.O.S. Printers Pte Ltd.
13 14 15 25 24 23

Contents

Geographic concepts and skills

The aim of standard Geography 1.4 is to test your ability to apply concepts and basic geographic skills to demonstrate understanding of a given environment. In essence, the standard assesses your ability to:

1 Use basic skills and geographic conventions in the presentation and/or interpretation of information

2 Show understanding of geography concepts.

At this point it is important to appreciate that the achievement standard places equal emphasis on your understanding of geographic concepts as it does your ability to apply basic geographic skills. In fact, such is the importance of both aspects of the standard, they are both assessed to Excellence in Level 1 NCEA Geography (Figure 1).

	Achieved	**Merit**	**Excellence**
Using basic skills and geographic conventions	✓	✓	✓
Showing understanding of geography concepts	✓	✓	✓

Figure 1

It follows then that you should seek to demonstrate your understanding and application of geographic skills and concepts as you respond to the learning activities in this book, and ultimately the final examination.

Applying basic geographic skills and conventions

Geographic skills are assessed in the external examination standard Geography 1.4 and the internal research standard Geography 1.5.

In the external examination you will be provided with a resource booklet, which you will use to show your understanding and application of geographic skills and concepts. The booklet will include a variety of resources such as maps, tables, diagrams, photographs and opinions. The resources in the booklet will be unfamiliar to you but will generally be about a particular geographic issue, which could be from New Zealand or from an overseas setting.

ISBN: 9780170368155

Basic geographic skills

Visual

- Interpretation of photographs, cartoons or diagrams including age-sex pyramids and models such as a wind rose
- Interpreting and completing a continuum to show value positions

Mapping

- Use of six-figure grid references and latitude and longitude
- Compass direction and bearings
- Distance, scale, area calculation
- Location of natural and cultural features
- Determination of height, cross-sections
- Use of a key, précis map construction
- Recognition of relationships, application of concepts
- Interpretation of other geographic maps like weather maps, cartograms, choropleth maps

Graphing

- Interpretation and construction of bar graphs (single and multiple), line graphs (single and multiple), pie and percentage bar graphs, scatter graphs, dot distribution, pictograms and climate graphs

Tables

- Recognition of pattern
- Simple calculation such as mean, mode and conversion to percentages

Figure 2

Understanding the geographic concepts or the 'big ideas' of geography

As a student of NCEA Geography you will be required to gain an understanding of how certain concepts or 'big ideas' underpin the knowledge and skills necessary to interpret information about the world. Although not comprehensive, the list of concepts (below) forms the basis of *all* learning in NCEA Geography (Figure 3).

Several Maori concepts also have relevance to Geography including kaitiakitanga (to care for), taonga (physical or cultural resource) and hekenga (migration).

Environments

Environments may be natural and/or cultural. They have particular characteristics and features which can be the result of natural and/or cultural processes. The particular characteristics of an environment may be similar to and/or different from another.

Perspectives

Perspectives are ways of seeing the world that help explain differences in decisions about, responses to, and interactions with environments. Perspectives are bodies of thought, theories or world-views that shape people's values and have built up over time. They involve people's perceptions (how they view and interpret environments) and viewpoints (what they think) about geographic issues. Perceptions and viewpoints are influenced by people's values (deeply held beliefs about what is important or desirable).

ISBN: 9780170368155

Processes

A process is a sequence of actions, natural and/or cultural, that shape and change environments, places and societies. Some examples of geographic processes include erosion, migration, desertification and globalisation.

Patterns

Patterns may be spatial: the arrangement of features on the earth's surface; or temporal: how characteristics differ over time in recognisable ways.

Interaction

Interaction involves elements of an environment affecting each other and being linked together. Interaction incorporates movement, flows, connections, links and interrelationships. Landscapes are the visible outcome of interactions. Interaction can bring about environmental change.

Change

Change involves any alteration to the natural or cultural environment. Change can be spatial and/or temporal. Change is a normal process in both natural and cultural environments. It occurs at varying rates, at different times and in different places. Some changes are predictable, recurrent or cyclic, while others are unpredictable or erratic. Change can bring about further change.

Sustainability

Sustainability involves adopting ways of thinking and behaving that allow individuals, groups and societies to meet their needs and aspirations without preventing future generations from meeting theirs. Sustainable interaction with the environment may be achieved by preventing, limiting, minimising or correcting environmental damage to water, air and soil, as well as considering ecosystems and problems related to waste, noise, and visual pollution.

Figure 3

Examination ready

As with the final examination, most questions in this book include a *command word*. Recognising the command word and understanding its requirements are essential to answering it successfully (Figure 4). For example, an examination question that asks you to explain the natural process that produced an extreme natural event, would require you to give a detailed account of the sequence of actions that caused the extreme natural event to occur.

ISBN: 9780170368155

Command word	Meaning	Example question
Annotate	Add brief notes to a diagram or graph.	Draw an annotated diagram to show the processes involved in an extreme natural event.
Classify	Arrange or order by class or category.	Classify these resources into renewable or non-renewable.
Compare	Give an account of the similarities between two or more items or situations, referring to all of them throughout.	Compare the effects of an ageing population to that of a youthful one.
Construct	Display information in a diagrammatic or logical form.	Construct a multi-line graph to illustrate the data in the table.
Define	Give the precise meaning of a word, phrase, concept or physical quantity.	Define the Maori concept of 'kaitiakitanga'.
Describe	Give a detailed picture of a given situation, event, pattern, trend, process or feature.	Describe the trend in population growth shown on the graph.
Discuss	Give a considered and balanced review that includes a range of arguments or factors. Opinions or conclusions should be presented clearly and supported by examples.	'Floods are more hazardous than earthquakes.' Discuss this statement.
Draw	Represent using a labelled, accurate diagram or graph using a pencil and a ruler. Diagrams should be drawn to scale, and graphs should have labelled axes, correctly plotted (if relevant) points joined in a line.	Draw a diagram to illustrate the natural elements of an environment interact.
Estimate	Give an approximate value.	Estimate the changes in the life expectancy rate over time.
Explain	Give a detailed account including reasons or causes.	Explain the relationship between fertility and female literacy.
Identify	Give an answer from a number of possibilities.	Identify the year in which there is an anomaly on the graph.
Label	Add labels to a diagram.	Label the main features on the diagram.
Outline	Give a brief account or summary.	Outline the impact of the One Child Policy.
State	Give a specific name, value or other brief answer — no need for explanation or calculation.	State the name of the cultural feature shown in the foreground of the photograph.

Figure 4

As will be the case in the final examination, you are advised to use coloured pencils (black, blue, green, red, brown, and yellow), a calculator, and a ruler where appropriate when answering the questions in this book. You should use coloured pencils when constructing diagrams and maps. However, labels and annotations on these diagrams and maps must be in pen. Also note that in the final examination, written work done in pencil will not be eligible for reconsideration.

The chapters that follow will offer you the opportunity to practise the types of questions you can expect in the final examination. Within this book, you will be required to apply concepts and basic skills using the specific resources provided, as you would in the final examination; the only difference being that in the examination, the resources will be presented in a separate booklet from that which contains the question and your answers.

ISBN: 9780170368155

Map interpretion and construction

A map is a two-dimensional graphic representation of a region of the earth's surface. Knowing how to interpret a map is a very important skill for geographers to have, as they provide us with information about places and help us to identify patterns and changes in a landscape.

The amount of detail contained within a map will vary according to its purpose and scale. However, most maps will have the following elements in common:

- a title stating the location and purpose of the map
- direction of orientation (i.e. north point)
- a scale
- a key or legend
- a border
- an indication of location (e.g. latitude and longitude).

There are many different types of map, each varying according to their purpose. The most common map type used by geographers is the *topographic* map. Topographic maps are detailed, accurate graphic presentations of features on the earth's surface. Examples of features found on topographic maps include:

Cultural	roads, buildings, urban areas, transport networks and place names
Drainage	rivers, lakes, streams and swamps
Relief	mountains, valleys, slopes and depressions
Vegetation	native and plantation forest, vineyards and orchards

Table 1.1 Features found on topographic maps

To read and use topographic maps, a geographer needs to be able to interpret:

- relative and absolute location
- latitude and longitude
- six-figure grid references
- compass points and bearings
- scale
- cross-sections
- map symbols
- précis maps.

The main topographic map series for New Zealand is the Topo50 and the Topo250 series, produced and published by Land Information New Zealand (LINZ). Topo50 series maps are produced to a scale of 1:50 000, and Topo250 series maps are produced to a scale of 1:250 000. Both show geographic features in detail, making them ideal for studies of the environment (Resource 1.1).

ISBN: 9780170368155

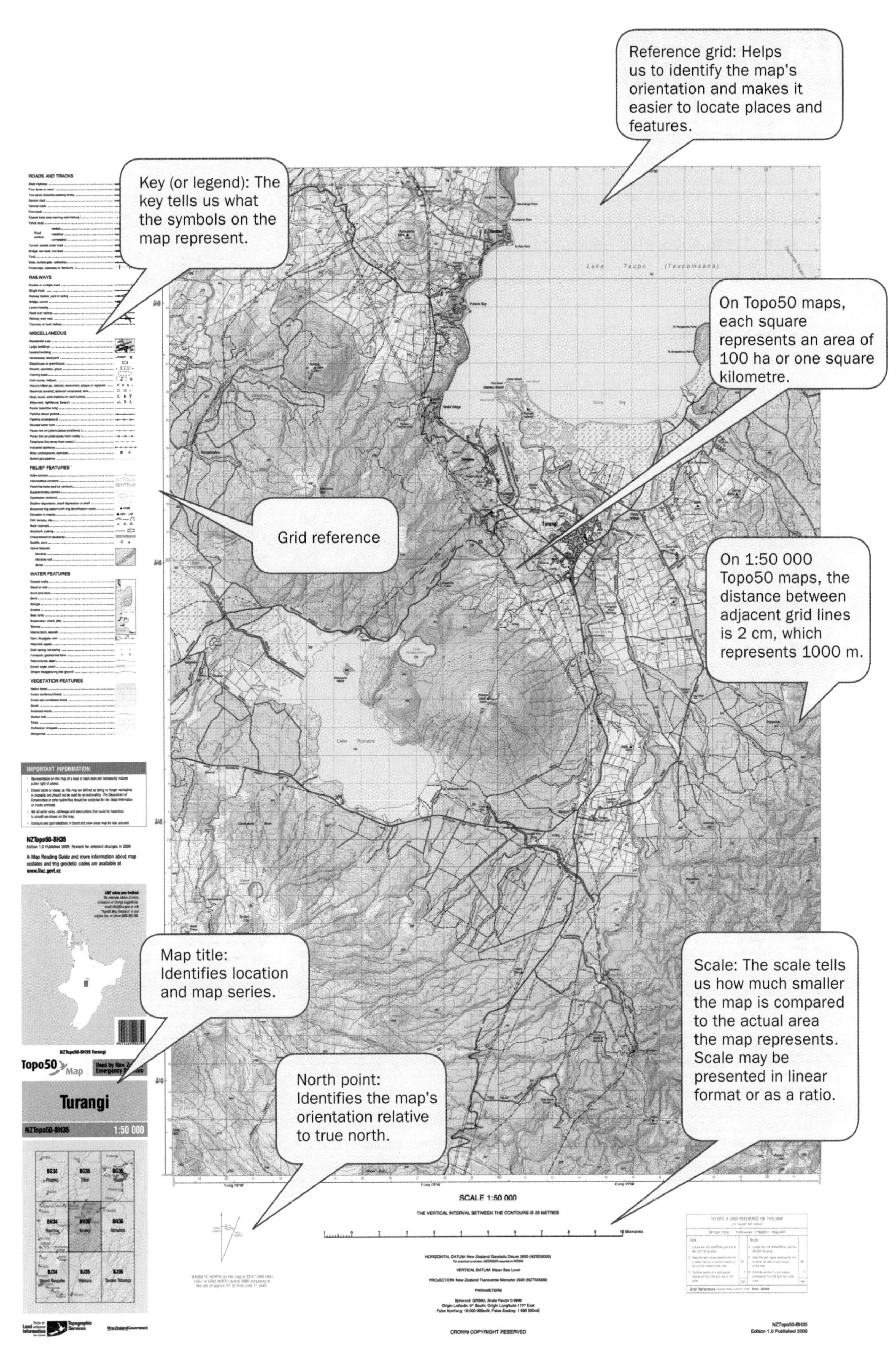

Resource 1.1 Topo50 map

ISBN: 9780170368155

1 Map symbols

Symbols on topographic maps show the location of features on the earth's surface. Therefore, to understand a map you need to understand its symbols.

Many symbols used on a topographic map look like the features they represent however some do not. This is why it is common practice for a topographic map to have a key or legend. The purpose of a key or legend is to show what each symbol actually means.

As illustrated in Resource 1.2, symbols used on a topographic map are grouped in the key under six broad headings:

- roads and tracks
- railways
- miscellaneous features of the built environment
- relief features
- water features
- vegetation features.

You will also notice from Resource 1.2 that colour plays an important part in symbol design and because of this some international conventions apply to the use of colour. For example, blue is used for water features, black for cultural features and green for vegetation.

To read map symbols correctly, follow the steps below:

1 Select a portion of the topographic map you wish to study.

2 Identify a symbol used in the area.

3 Find the symbol in the key and identify what the symbol represents.

4 Repeat steps 2 and 3 until you have identified all of the major symbols used in the area.

Try doing this in pairs with the map found in Resource 1.1.

ROADS AND TRACKS [1]

State highway
Four lanes or more
Two lanes (includes passing lanes)
Narrow road
Vehicle track
Foot track
Closed track (see warning note below)[2]
Poled route
Road surface: sealed, metalled, unmetalled
Tunnel, tunnel under road
Bridge; two lane, one lane
Ford
Gate, locked gate, cattlestop
Footbridge, cableway or handwire [3]

RAILWAYS

Double or multiple track
Single track
Railway station, yard or siding
Bridge, tunnel
Level crossing
Road over railway
Railway over road
Tramway or bush railway

MISCELLANEOUS

Residential area
Large buildings
Isolated building
Homestead, stockyard — Awapuni
Glasshouse or greenhouse
Church, cemetery, grave
Training track
Golf course, helipad
Historic Māori pa, redoubt, monument, plaque or signpost
Reservoir covered, reservoir uncovered, tank
Mast, tower, wind machine or wind turbine
Shipwreck, lighthouse, beacon
Fence (selection only)
Pipeline above ground
Pipeline underground
Disused water race
Power line on pylons (actual positions) [3]
Power line on poles (away from roads) [3]
Telephone line (away from roads) [3]
Industrial cableway
Mine; underground, opencast
Buried gas pipeline

RELIEF FEATURES [4]

Index contour — 100
Intermediate contours
Perennial snow and ice contours
Supplementary contour
Depression contours
Shallow depression, small depression or shaft
Beaconed trig station (with trig identification code) — A1B2
Elevation in metres — 130m · 130
Cliff, terrace, slip
Rock outcrops
Stopbank, cutting
Embankment or causeway
Saddle, cave
Alpine features
Moraine
Moraine wall
Scree

WATER FEATURES

Coastal rocks
Shoal or reef
Sand and mud
Sand
Shingle
Swamp
Boat ramp
Breakwater, wharf, jetty
Slipway
Marine farm, seawall
Dam, floodgate, weir — FG
Waterfall, rapids
Cold spring, hot spring
Fumarole, geothermal bore
Watercourse, drain
Canal; large, small
Stream disappearing into ground

VEGETATION FEATURES

Native forest
Exotic coniferous forest
Exotic non-coniferous forest
Scrub
Scattered scrub
Shelter belt
Trees
Orchard or vineyard
Mangroves

Resource 1.2

ISBN: 9780170368155

Learning Activities

1 Use the key in Resource 1.2 to complete the following task.

a Draw the symbol used to show each of the following features:

State highway	
Bridge	
Tunnel	
Single railway track	
Church	
Cemetery	
Historic Maori pa	
Shipwreck	
Underground mine	
Spot elevation	
Cliff	
Waterfall	
Exotic forest	
Orchard	

b Identify the feature represented by the following symbols:

100	
×	

2 Locating places: direction and bearing

The notion of location (or the particular place where something is) is the most basic of all geographical concepts. It is therefore important that, as a student of geography, you refine the skill of locating places before you master any other.

Location is usually expressed in one of two ways:

- **Relative location** describes the location of a place or feature as it relates to other features.
- **Absolute location** refers to the location of a point on the earth's surface, referenced to a grid system (as it appears on a map). Of the many grid referencing systems in use, the two that are important for students of Level 1 NCEA Geography is the six-figure grid reference system, and latitude and longitude.

Relative location is usually expressed in one of two ways. The most common method is to describe *direction* according to the points of a compass; however it can also be expressed as a *bearing* when a more precise reading is required.

Resource 2.1

The four *cardinal* (main) points of the compass are north (N), south (S), east (E) and west (W). These points in turn can be divided even further with *intermediate* points to form the eight-point compass rose (Resource 2.1).

ISBN: 9780170368155

As a compass is a circle made up of 360 degrees, its directions can also be expressed as an angle or bearing. Bearings can be measured on a map with the aid of a protractor. Bearings are always stated as three-figures as measured clockwise from north (000°). For example, east is 090° clockwise of north, south-east 135° and north-west 315° clockwise of north (Resource 2.2).

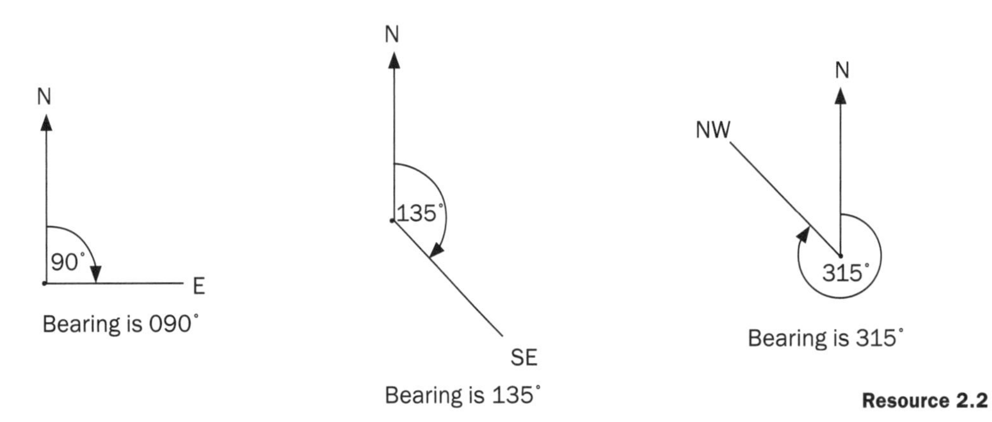

Resource 2.2

Of course, to measure bearing on a map, you first need to establish the direction of north. Most topographic maps will show *grid north* in the form of parallel north-south grid lines. Grid north differs from magnetic and true north in that it provides a fixed orientation to measure from. Most maps will show grid north (000°) pointing to the top of the page, however exceptions to this rule are not uncommon, so always confirm the map's orientation before undertaking any measures.

To use bearings to show direction, follow the steps below:

1. Using a ruler and pencil, rule a line joining the points *x* and *y*.
2. Place a clear protractor over the line *x-y* so that the 0° reading on the protractor points to grid north and the *x* point is positioned at the centre of the protractor.
3. Reading clockwise from 0°, state the angle the *x-y* line intersects the measuring edge of the protractor (Resource 2.3). The angle measured is the bearing from point *x* to point *y*.

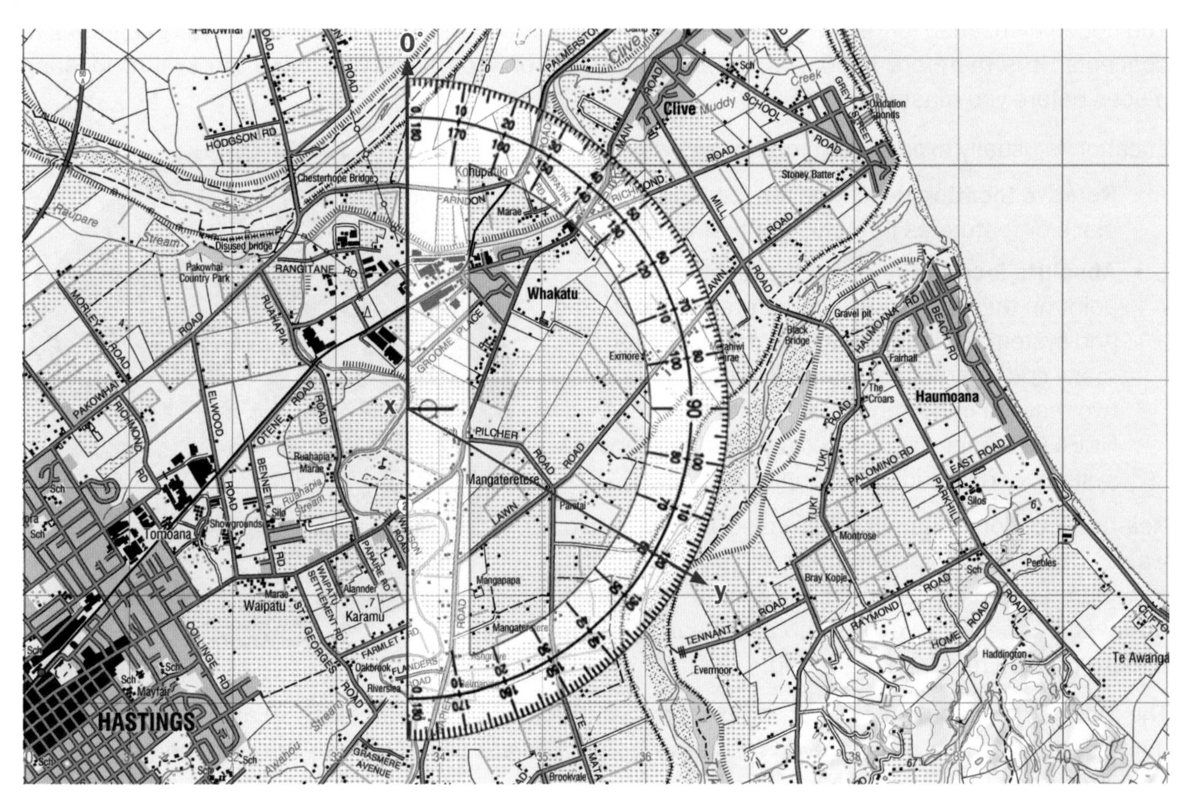

Resource 2.3

ISBN: 9780170368155

It is often useful to be able to describe the relative location of features on a map by expressing their location in terms of compass quadrants. This is done by dividing the map into quarters and then naming each quarter according to the points of a compass (Resource 2.4).

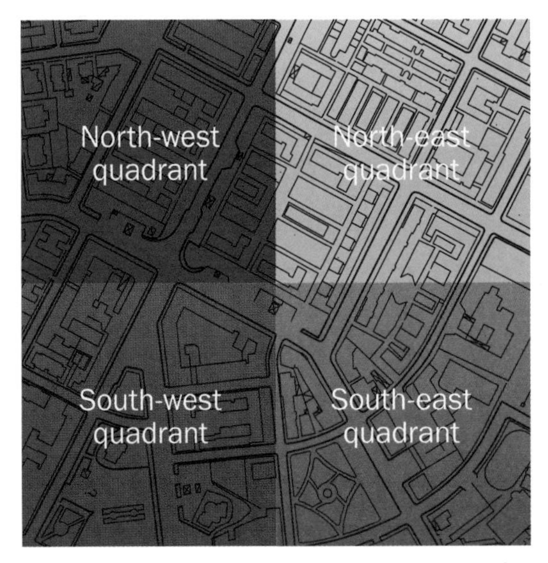

Resource 2.4

Learning Activities

1 Compass direction is often used to identify the regions of a country. With the aid of an atlas, locate and label the following regions of New Zealand on Resource 2.5.

a Northland

b East Cape

c Westland

d Southland

Resource 2.5

ISBN: 9780170368155

Learning Activities

2 Direction is also used to describe the regions of a continent. With the aid of an atlas, locate and shade the following regions of Asia on Resource 2.6:

a East Asia **b** South-east Asia **c** North Asia

d North-east Asia **e** Western Asia.

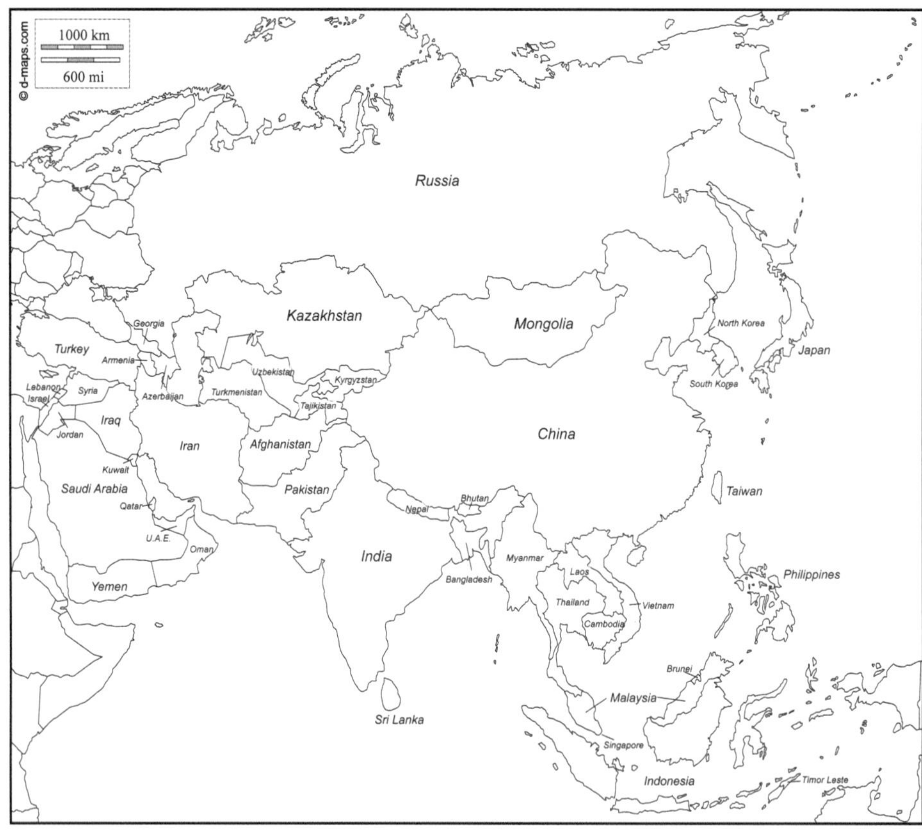

Resource 2.6

3 Study Resource 2.3 and then complete the following activities.

a State the compass direction from:

i Clive to Haumoana ______________________

ii Whakatu to Clive ______________________

iii Haumoana to Whakatu ______________________

b State the bearing direction from:

i Haumoana to Clive ______________________

ii Clive to Whakatu ______________________

iii Whakatu to Haumoana ______________________

4 Explain what the term *quadrant* means.

____

____

____

ISBN: 9780170368155

Learning Activities

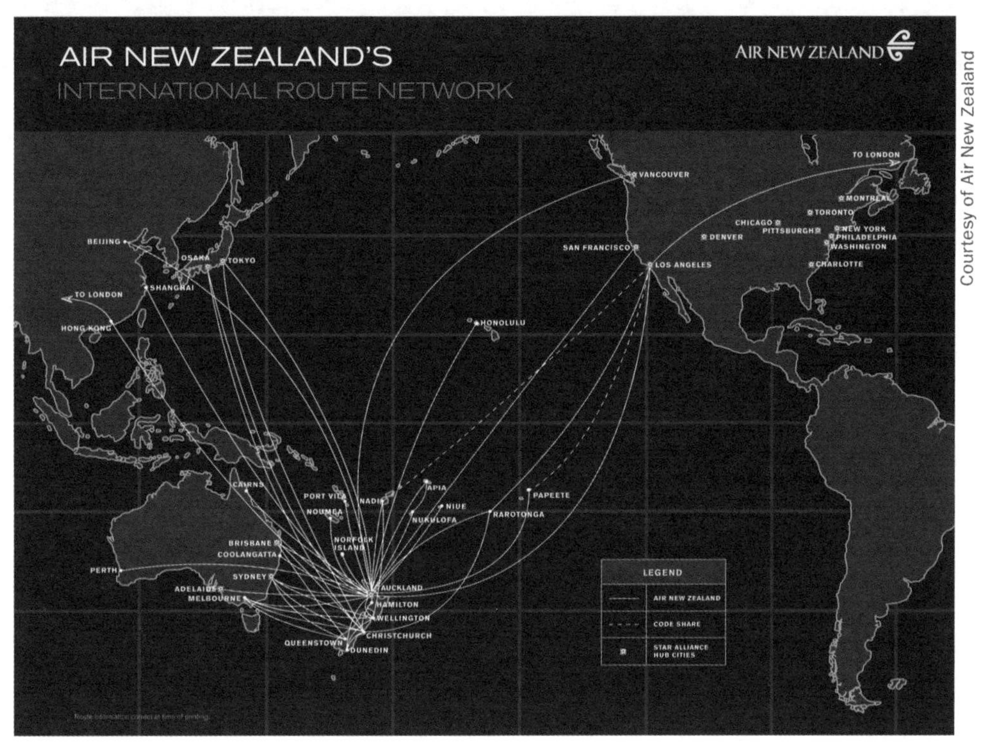

Courtesy of Air New Zealand

Resource 2.7

5 Study Resource 2.7 and complete the following activities.

a State the compass direction for flights travelling from:

i Auckland to San Francisco ____________________

ii Queenstown to Sydney ____________________

iii Rarotonga to Christchurch ____________________

iv Tokyo to Auckland ____________________

b State the bearing direction for flights travelling from:

i Auckland to San Francisco ____________________

ii Queenstown to Sydney ____________________

iii Rarotonga to Christchurch ____________________

iv Tokyo to Auckland ____________________

c Name three destinations Air New Zealand flies to in:

i The north-west quadrant of the map.

____

____

____

ii The south-west quadrant of the map.

____

____

____

ISBN: 9780170368155

3 Locating places: six-figure grid references

Maps are particularly useful to geographers when it comes to finding absolute location. The exact location of a feature on a topographic map, for example, is determined with *grid lines*. Grid lines are equally spaced vertical and horizontal lines drawn on a map and are called:

- **Eastings**: vertical grid lines that divide the map into columns from west to east (*x*-coordinate)
- **Northings**: horizontal grid lines that divide the map into rows from north to south (*y*-coordinate).

Eastings and northings intersect each other on a map to form grid squares. The grid squares are then used to help calculate unique six-figure grid references for individual features (Resource 3.1).

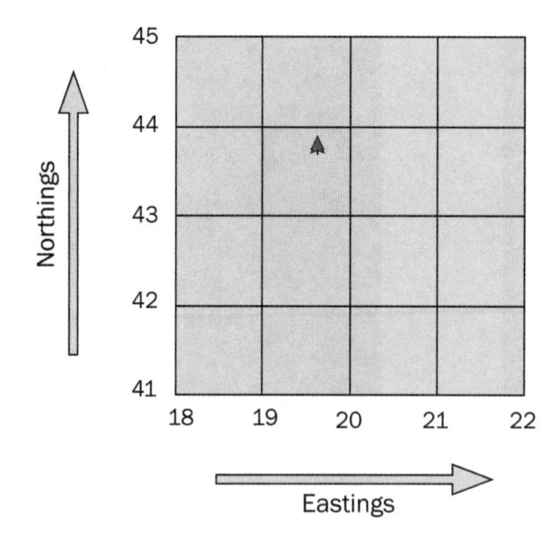

Resource 3.1

Six-figure grid references contain six digits. The first three digits of a six-figure grid reference refer to the easting, and the last three digits refer to the northing. The third and sixth digit is determined by dividing the easting and northing into tenths (Resource 3.2).

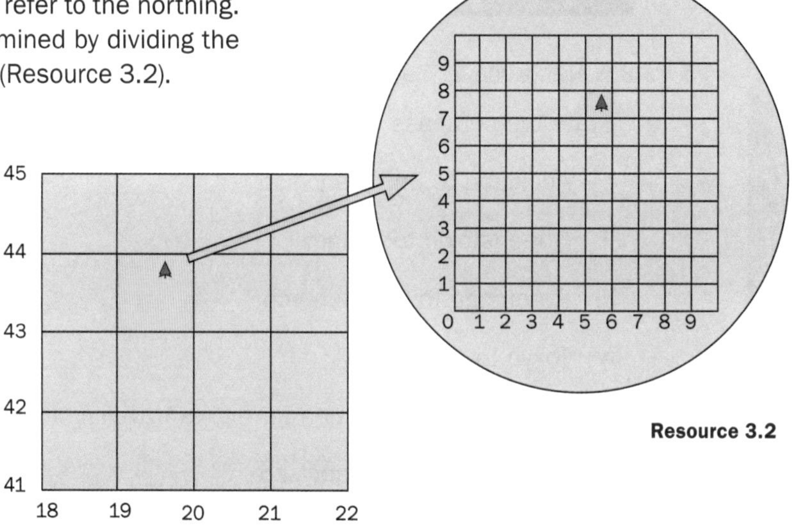

Resource 3.2

To give the exact absolute location of a feature using six-figure grid references, follow the steps below:

1. To determine the six-figure grid reference for the ▲ symbol in Resource 3.2, identify the first vertical grid line to the left of the symbol. In the example above, the first grid line to the left is labelled *19* in the bottom margin. This is the easting.
2. Estimate the number of tenths eastward the symbol is from the easting on the left. In the example above, the symbol is located a distance of $\frac{6}{10}$ from the easting on the left. This gives a final easting reading of *196*.
3. Now identify the first horizontal grid line below the symbol. In the example above, the first grid line below is labelled *43* in the left margin. This is the northing.
4. Estimate the number of tenths northward the symbol is from the northing beneath. In the example above, the symbol is located a distance of $\frac{7}{10}$ from the northing beneath. This gives a final northing reading of *437*.
5. You have now calculated the six-figure grid reference for the ▲ symbol. It should be written as GR 196437.

ISBN: 9780170368155

Learning Activities

1 Study Resource 3.3 and then complete the following task.

a State the six-figure grid reference for:

i Plumpton Dairy Farm ____________________

ii Village Hall ____________________

iii Grove Farm ____________________

iv The church (✝) ____________________

v The intersection of Grove Road and Dale Road ____________________

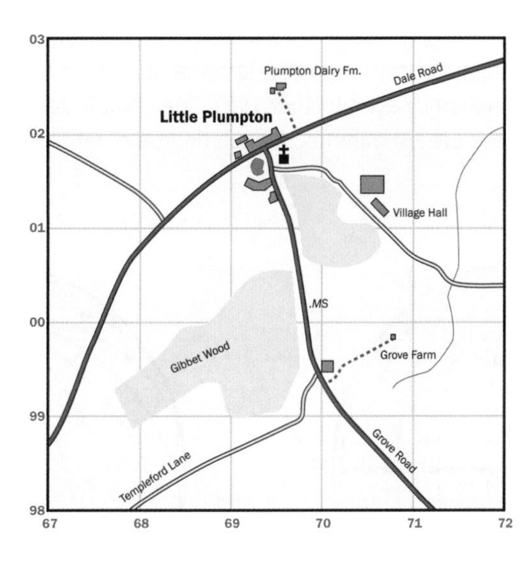

Resource 3.3

2 Study Resource 3.4 and then complete the following task.

a State the feature located at the following grid references:

i GR 664578 ____________________

ii GR 657591 ____________________

iii GR 688613 ____________________

iv GR 673580 ____________________

v GR 685565 ____________________

vi GR 664613 ____________________

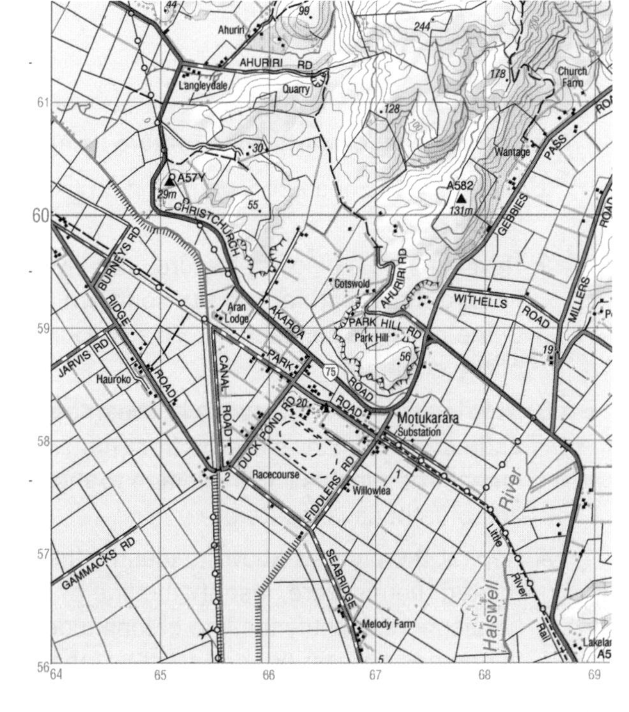

Resource 3.4

ISBN: 9780170368155

4 Locating places: latitude and longitude

The second method for determining absolute location is to use latitude and longitude.

Lines of latitude are imaginary lines that run east to west around the earth's circumference. They run parallel to each other and for this reason are also known as *parallels of latitude*. Latitude is measured in degrees north (N) or south (S) in relation to the equator, which itself represents zero degrees (0˚). The equator divides the earth into the northern hemisphere and southern hemisphere. The latitude of the north pole is 90˚N of the equator while the latitude of the south pole is 90˚S of the equator.

Lines of longitude run in a north to south direction from the north to the south pole, and on a two-dimensional map intersect lines of latitude at right angles. Longitude is measured in degrees west (W) or east (E) of the *prime meridian* (0˚). The prime meridian divides the Earth into the western and eastern hemispheres.

Together, lines of latitude and longitude form a grid system that allows us to pinpoint places and features on the earth's surface. To increase accuracy even further, each line of latitude and longitude can be divided into smaller units called *minutes*. There are 60 minutes in each degree of latitude or longitude.

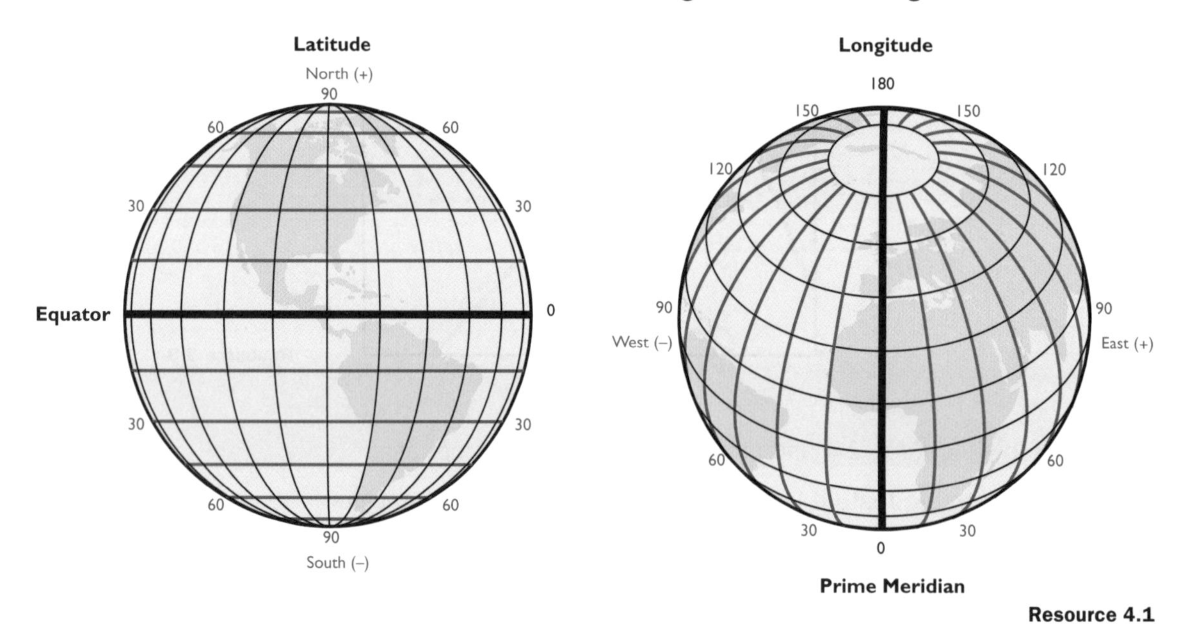

Resource 4.1

To state the exact location of a feature using latitude and longitude, follow the steps below:

1. Using a world or country map, find a place or feature you wish to calculate latitude and longitude for.
2. Having found a place or feature, identify the line of latitude directly to the north of the feature if the feature is in the southern hemisphere, or directly to the south if the feature is in the northern hemisphere. This is your line of latitude in degrees (°) north (N) or south (S) of the equator.
3. Now imagine that the space between each line of latitude is divided into 60 minutes. Estimate how many sixtieths from the line of latitude the feature is located. This is the number of minutes ('), in addition to the degrees north or south of the equator, that your line of latitude is located.
4. Repeat the process above to identify the line of longitude. If your feature is located in the eastern hemisphere, identify the line of longitude directly to the west. If it is in the western hemisphere, identify the line of longitude directly to the east. This is your line of longitude in degrees (°) west (W) or east (E) of the prime meridian.
5. Now estimate the number of minutes as above.
6. Remember to always state latitude first and longitude last.

ISBN: 9780170368155

As an example, Resource 4.2 shows that Wellington has an approximate latitude of 41°S and a longitude of 174°E. To give a more precise reading using both degrees and minutes, however, the location of Wellington could be expressed as 41°17'S 174°47'E, which tells us it is 41 degrees and 17 minutes south of the equator, and 174 degrees and 47 minutes east of the prime meridian.

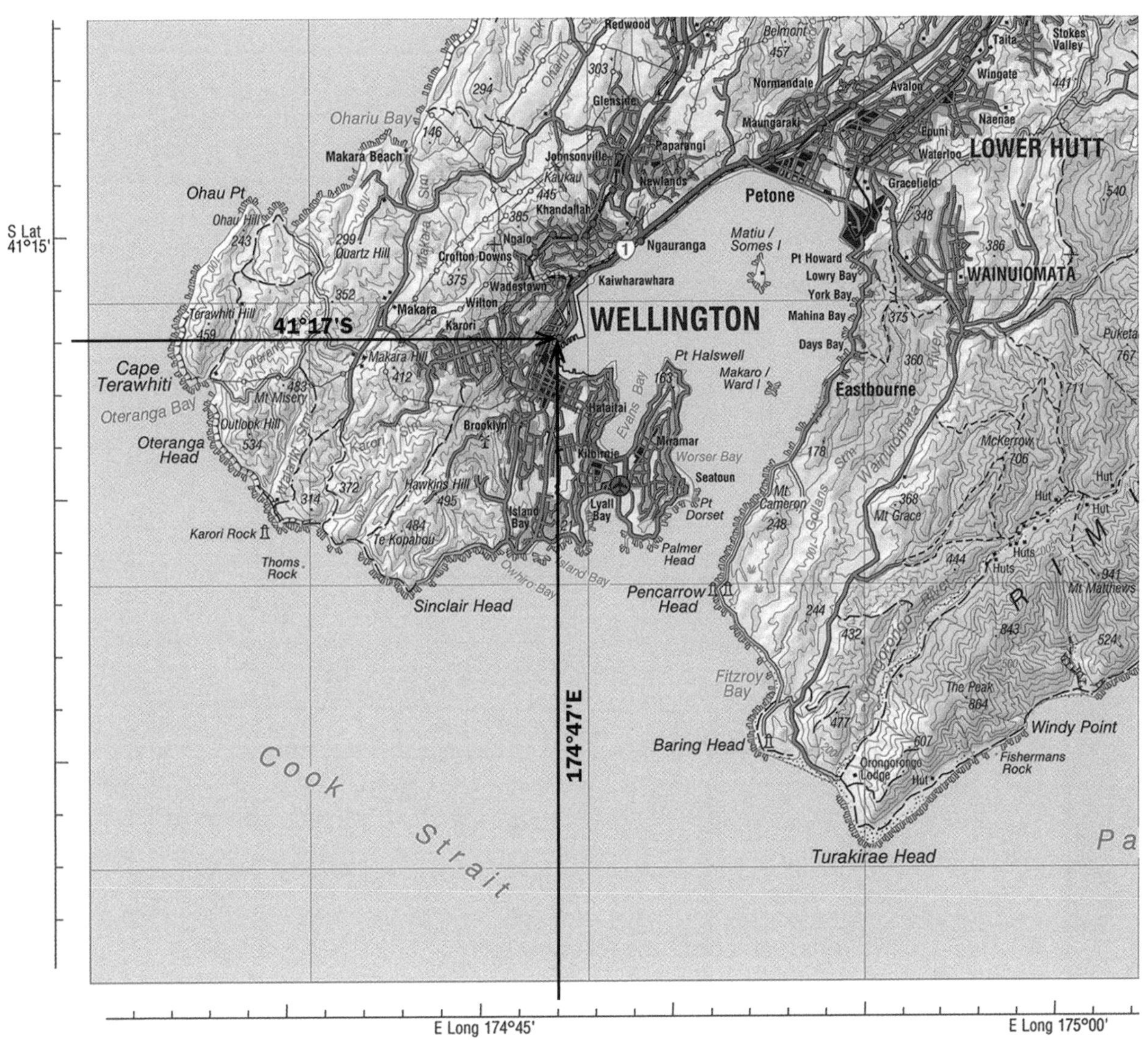

Resource 4.2

Learning Activities

1 Describe the difference between parallels of latitude and meridians of longitude.

ISBN: 9780170368155

Learning Activities

2 Describe the location and significance of the:

a Equator ______________________________

____

____

b Prime meridian ______________________________

____

____

c North and South poles ______________________________

____

____

____

d Tropic of Cancer ______________________________

____

____

____

e Tropic of Capricorn ______________________________

____

____

3 With the aid of an atlas, complete the following table.

Brisbane, Australia	
	23°43'N 90°25'E
	49°16'N 123°07'W
London, England	
Berlin, Germany	
	64°09'N 21°51'W
	17°58'N 76°48'W
Tokyo, Japan	
Auckland, New Zealand	
	25°15'N 51°36'E

ISBN: 9780170368155

Learning Activities

4 Read the following account of Captain James Cook's first voyage to the South Pacific Ocean aboard HMS *Endeavour*, which took place between 1768 and 1771.

The aims of the first expedition to the South Pacific Ocean were to observe the 1769 transit of Venus across the Sun, and to seek evidence of the unknown southern land. Departing Plymouth, UK in August 1768, the expedition crossed the Atlantic, rounded Cape Horn and reached Tahiti in time to observe the transit of Venus. In September 1769, the expedition reached New Zealand. Cook and his crew spent the following six months charting the New Zealand coast, before resuming their voyage westward across the Tasman Sea. In April 1770, Cook's crew became the first Europeans to reach the east coast of Australia, making landfall on the shore of what is now known as Botany Bay. The expedition continued northward along the Australian coastline, narrowly avoiding shipwreck on the Great Barrier Reef. In October 1770, the badly damaged *Endeavour* came into a port in Batavia (Jakarta) in the Dutch East Indies (Indonesia). They resumed their journey on 26 December, rounded the Cape of Good Hope on 13 March 1771, and reached the English port of Deal, Kent on 12 July. The voyage lasted nearly three years.

Captain Cook's first voyage 1768-1771

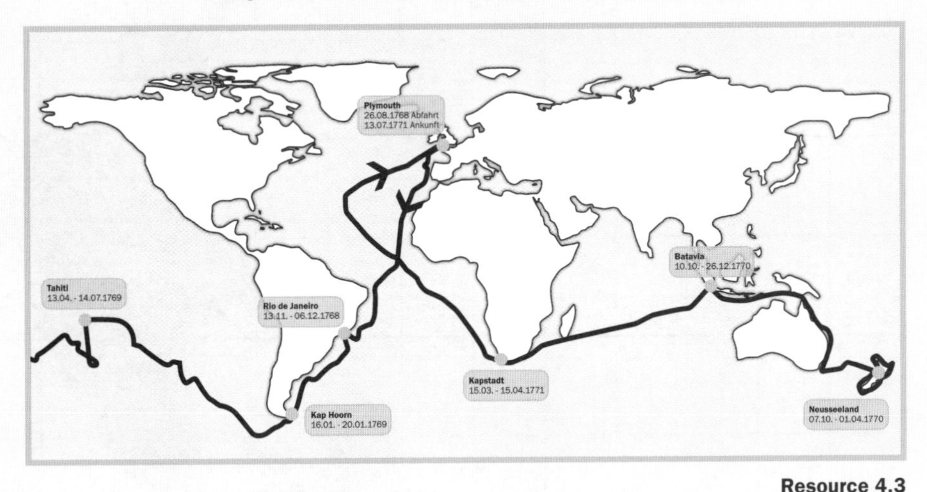

Resource 4.3

a With the aid of an atlas, estimate the latitude and longitude for each of the ports visited by Captain Cook during his first voyage to the South Pacific (Resource 4.3).

Plymouth, UK	
Rio de Janeiro, Brazil	
Cape Horn, Chile	
Tahiti	
New Zealand	
Botany Bay, Australia	
Batavia, Dutch East India	
Cape Town, South Africa	

ISBN: 9780170368155

5 Using the map in Resource 4.4, estimate the latitude and longitude of the following places:

a Auckland ______________________

b Tauranga ______________________

c Mahia Peninsula ______________________

d Mt Ruapehu ______________________

e Christchurch ______________________

f Dunedin ______________________

g Queenstown ______________________

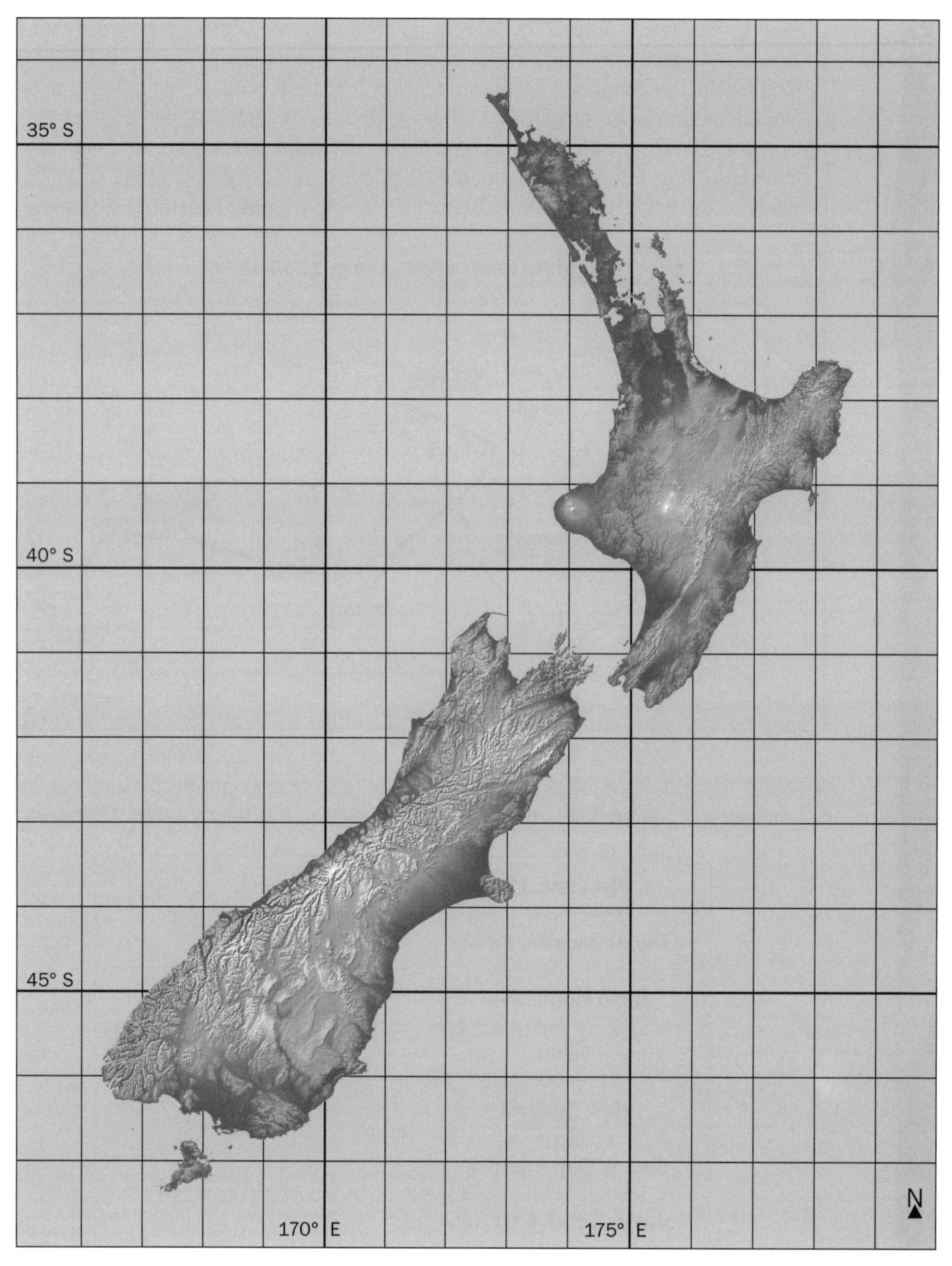

Resource 4.4

ISBN: 9780170368155

6 Did you know the world's time zones are based on lines of longitude? The world is divided into separate times zones each with a width of 15°; equal to the distance the sun appears to travel across the sky in any given hour. In total, there are 24 time zones, corresponding to the number of hours in a day (Resource 4.5).

Because the earth spins in an easterly direction, places to the east are ahead in time while places to the west are behind. For example, Wellington, New Zealand is two time zones ahead of Sydney, Australia. In other words, if it is 1 pm in Wellington it will be 11 am in Sydney at the same time!

Using the world time zone map in Resource 4.5, calculate how many hours Wellington is ahead or behind the following locations:

a Western Australia ______________________

b United Kingdom ______________________

c Japan ______________________

d India ______________________

e Chile ______________________

Standard Time Zones of the World

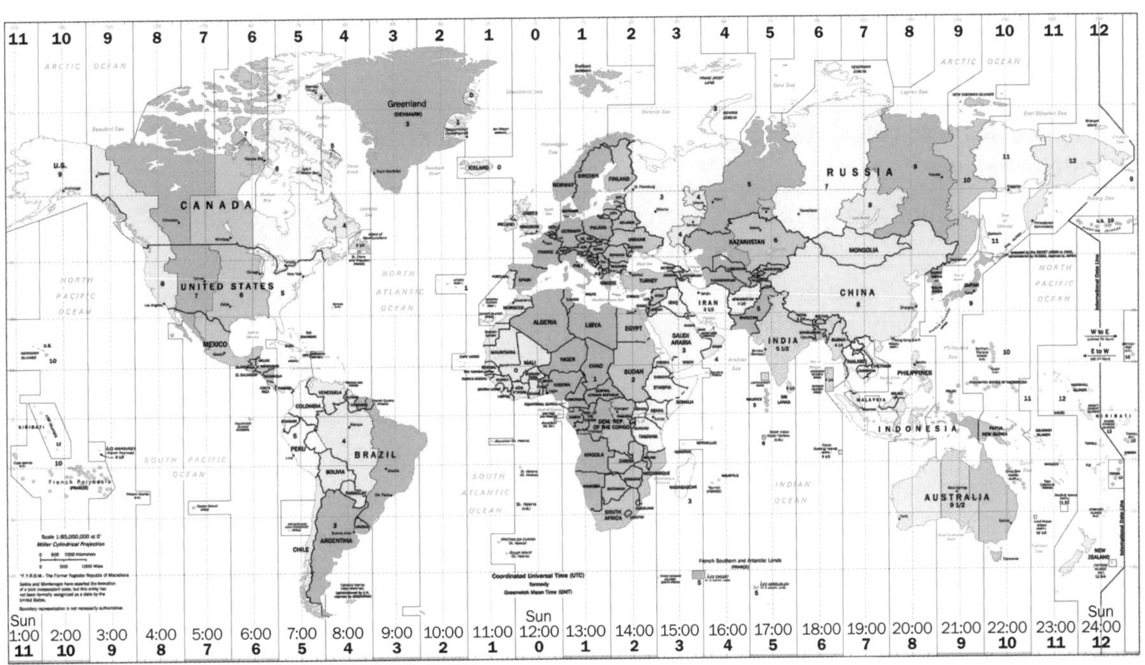

Resource 4.5

ISBN: 9780170368155

5 Using scale: calculating distance

A map represents an area on the earth's surface and a map's scale refers to the relationship between distances portrayed on a map and the distance on the ground. Put in another way, a map is a scaled down representation of part of the earth's surface. This means that a map's scale can be used to calculate surface distances.

There are three accepted ways of showing scale on a topographic map:

- As a **written statement**, for example, on the New Zealand Topo50 series 1 cm is equal to 50 000 cm (or 500 m).
- As **ratio or representative fraction**, for example the 1:50 000 ratio of the Topo50 series can be expressed as the fraction: $\frac{1}{50\,000}$. Here, the numerator represents the number of units on the map while the denominator represents the number of units that one unit on the map is equal to on the ground. In this example, $\frac{1}{50\,000}$ means that one unit on the map equals 50 000 units on the ground.
- As a **linear scale** or scale bar:

SCALE 1:50 000

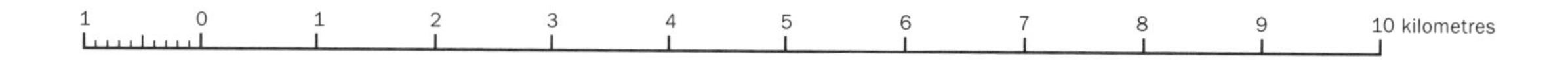

To use scale to calculate the distance between two points on a topographic map, follow the steps below:

1. To measure a straight line distance, place a ruler (or the edge of a sheet of paper) between the two points and measure the distance between them.
2. Next, place the ruler along the map's linear scale, overlaying the zero point on the ruler with the zero point on the linear scale.
3. Finally, read the distance of the second point off the linear scale.

To estimate distance along a curve (e.g. a river) replace the ruler with a piece of string and follow the steps above.

Learning Activities

1 Express the following scales as statements:

a 1:2 million ____________________

b 1:250 000 ____________________

c 1:50 000 ____________________

d $\frac{1}{2500}$ ____________________

e $\frac{1}{200}$ ____________________

ISBN: 9780170368155

Learning Activities

2 Express the following scales as representative fractions and ratios:

a 1 cm is equal to 500 m ______________________________

b 1 cm is equal to 2.5 km ______________________________

c 1 cm is equal to 10 km ______________________________

3 Refer to Resource 5.1. Measure the straight line distance of the following:

a Holder Lookout and Musick Point ______________________________

b Mt Wellington and Pigeon Mountain ______________________________

c The width of Panmure Basin at its widest point ______________________________

d Eastern Beach ______________________________

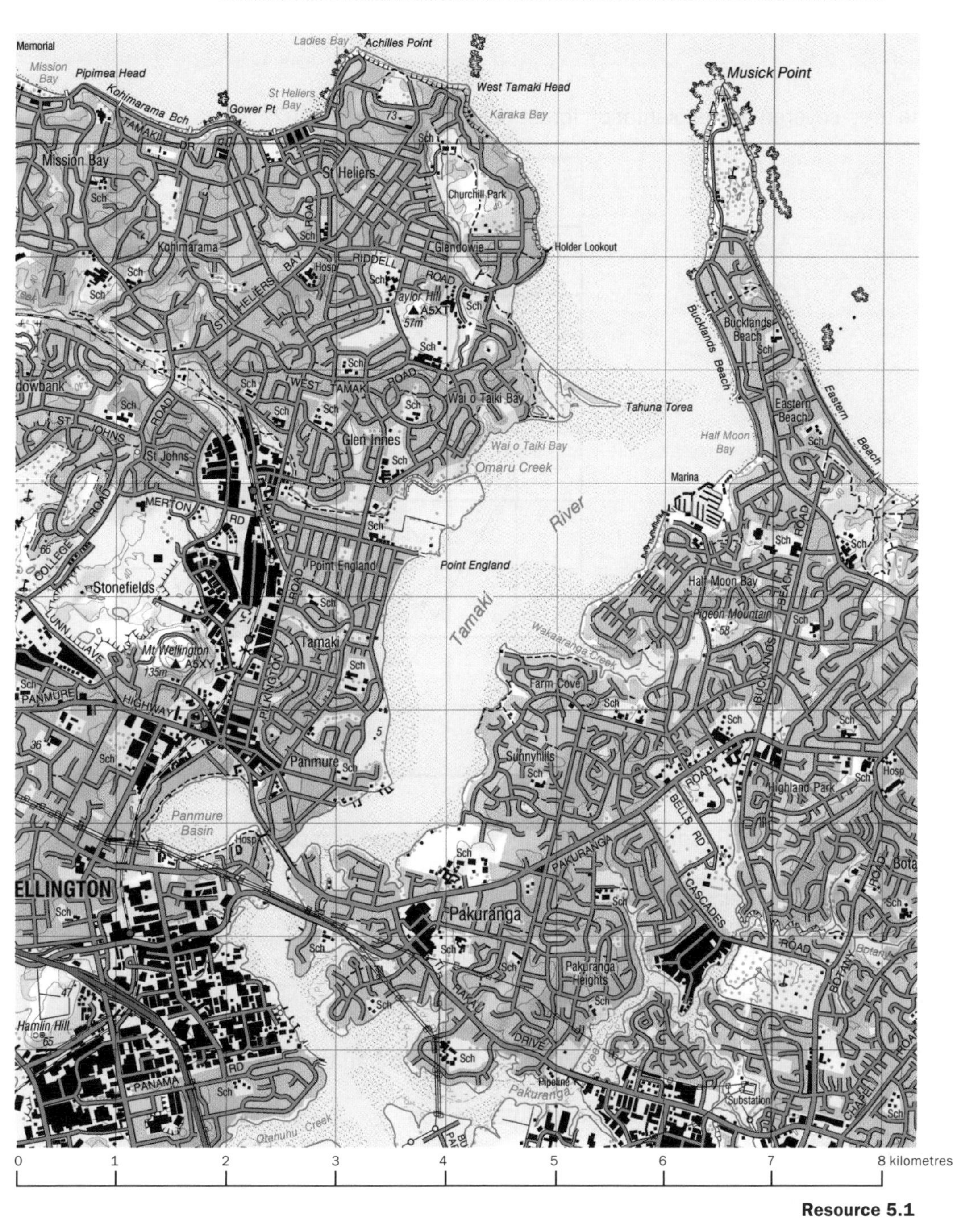

Resource 5.1

ISBN: 9780170368155

6 Using scale: estimating area

The area that a feature covers on the earth's surface can be calculated by using the scale of the map.

To estimate an area from a topographic map, follow the steps below:

1. Measure the distance of the map area from east to west in kilometres.
2. Measure the distance of the map area from north to south in kilometres.
3. i If the area to be measured is square or rectangular, multiply the east-west distance by the north-south distance.
   ii If the area to be measured is an irregular shape an accurate calculation will be difficult. However, an estimate of the area can be made by counting the number of grid squares covered by the feature. This can be done by counting the number of squares covered by more than half of the feature, and ignoring squares covered by less than half of the feature.
4. Express the area in square kilometres (km^2).

The area covered by the plantation forest below, for example, is 29 km^2.

= 1 km^2

✔ = 29

Area = 29 km^2

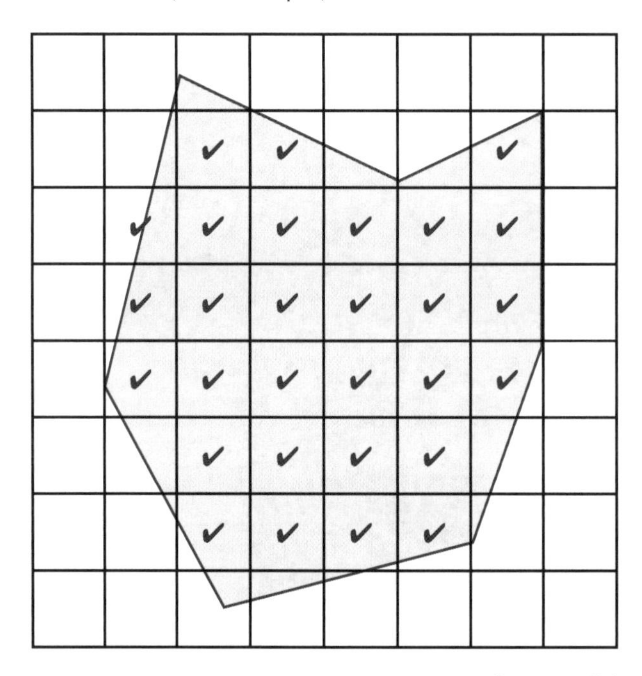

Resource 6.1

Learning Activities

1 Study Resource 5.1 on page 25 and then complete the following activities.

a Calculate the area bounded by the map ______

b Estimate the area covered by Tamaki River ______

c Estimate the built-up area of the map ______

d Estimate the area bounded by Ti Rakau Drive, Pakuranga Highway, Cascades Road and Botany Road ______

ISBN: 9780170368155

7 Showing and interpreting relief

Relief is a word used by geographers to describe the shape or pattern of landforms, such as its height above sea level or the steepness of its slopes.

Topographic maps show the pattern of relief in three ways:

- shading
- spot heights
- contour lines.

Of the three methods used, contour lines are most effective in showing relief patterns. *Contour lines* trace out areas of equal height or elevation above sea level, and give an indication of the *gradient* or slope of the land. Closely spaced contour lines indicate a steep gradient and contour lines spaced far apart indicate a gentle gradient (Resource 7.1).

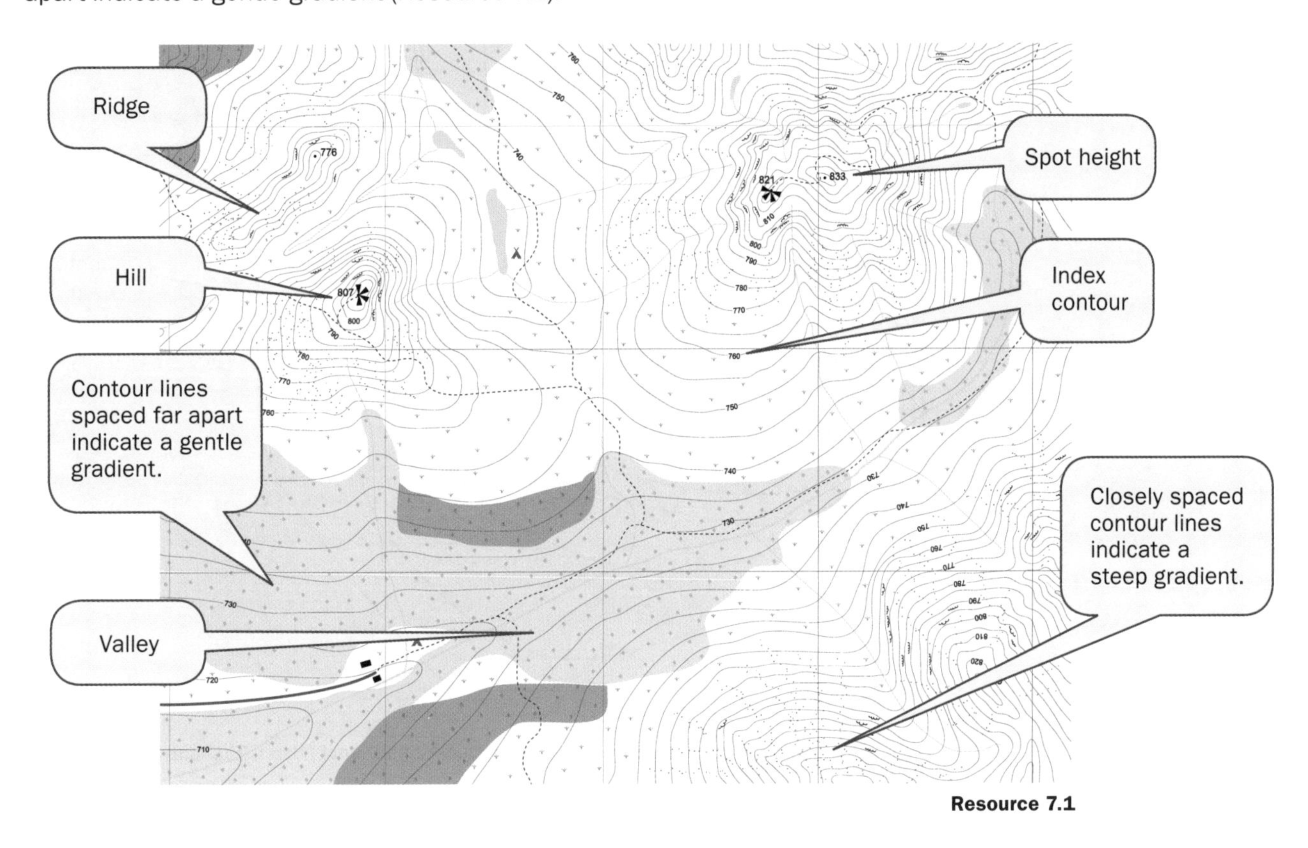

Resource 7.1

The vertical distance between adjacent contour lines is known as the *contour interval* (CI). The interval is always constant on any given map.

To use contour lines to determine the elevation of a feature on a topographic map, follow the steps below:

1. Find the contour interval of the map from the key or legend, and note both the interval and the unit of measure. New Zealand's Topo50 map series, for example, has a contour interval of 20 m.
2. Find the numbered contour line nearest the feature for which the elevation is being sought.
3. Determine the direction of slope from the numbered contour line to the desired feature.
4. Count the number of contour lines that must be crossed to go from the numbered line to the feature and note the direction up or down. The number of lines crossed multiplied by the contour interval is the distance above or below the starting value.
5. When the feature is on a contour line, its elevation is that of the contour. If the feature is between contour lines, then estimate the elevation to be half of the contour interval.

ISBN: 9780170368155

Cross-sections

Geographers routinely use information from contours to construct cross-sections. Cross-sections are an effective way to show the changing slope and shape of the land in profile. Resource 7.2 shows an example of how contours on a map can be used to construct a cross-section or profile.

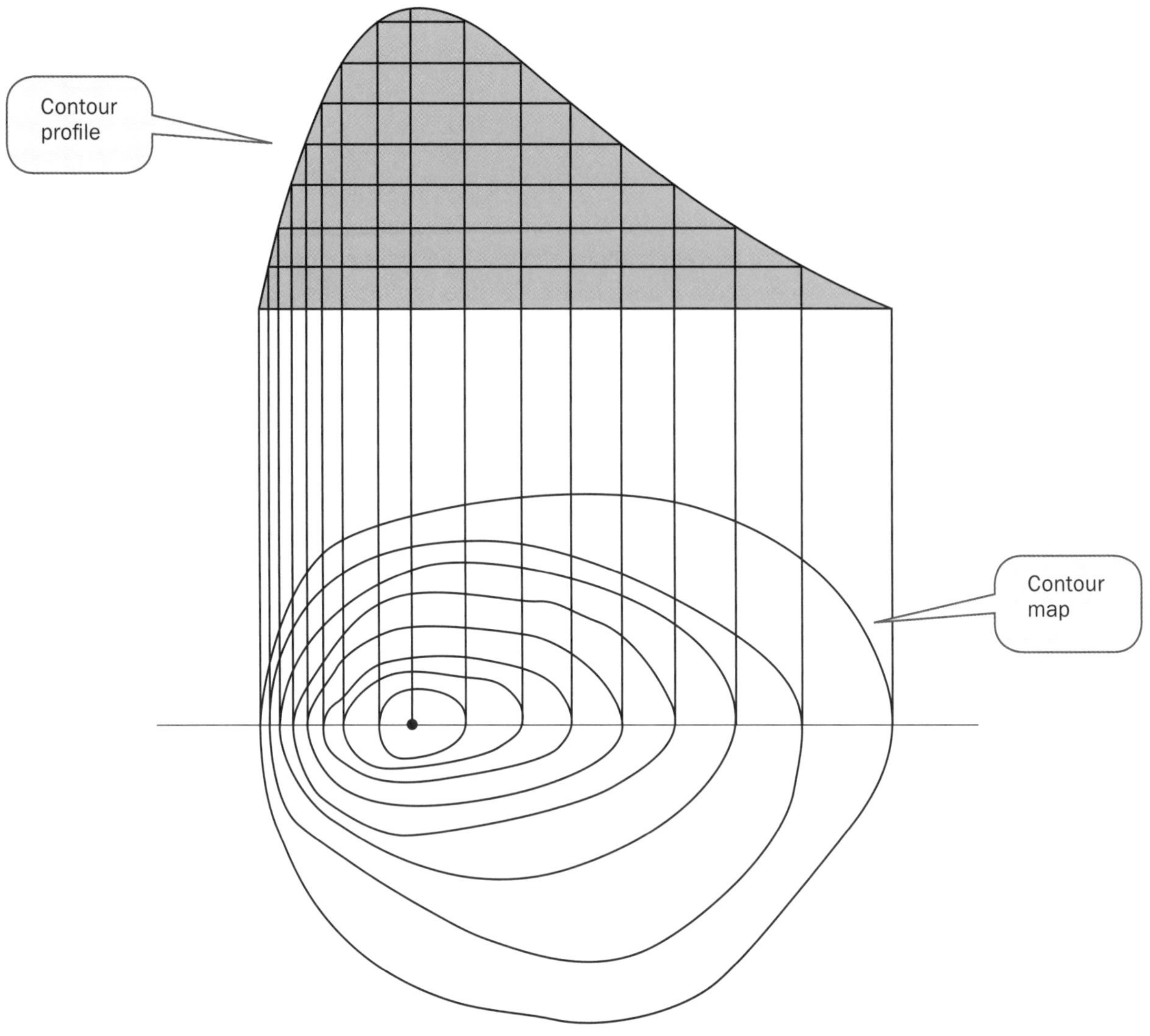

Resource 7.2 Contour map and cross-section of Ko Samui Island, Thailand

Ko Samui Island, Thailand

ISBN: 9780170368155

To construct a cross-section, follow the steps below:

1 Decide where you want the cross-section line to be.

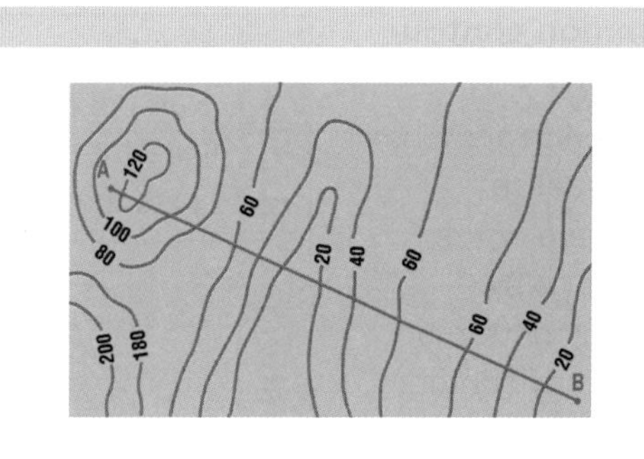

2 Place the edge of a sheet of paper along the line that joins points A and B. Mark points A and B on the edge of the sheet of paper.

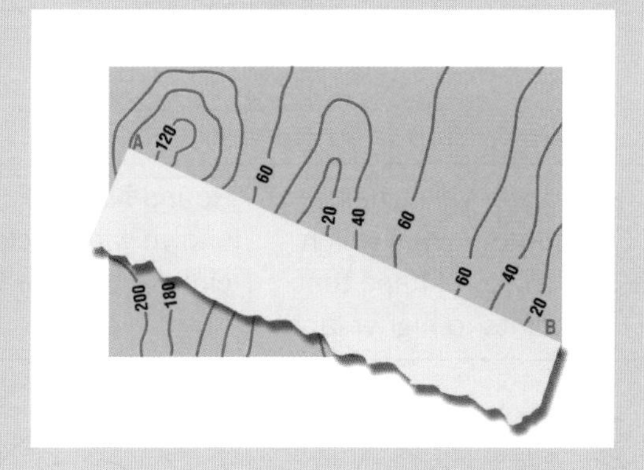

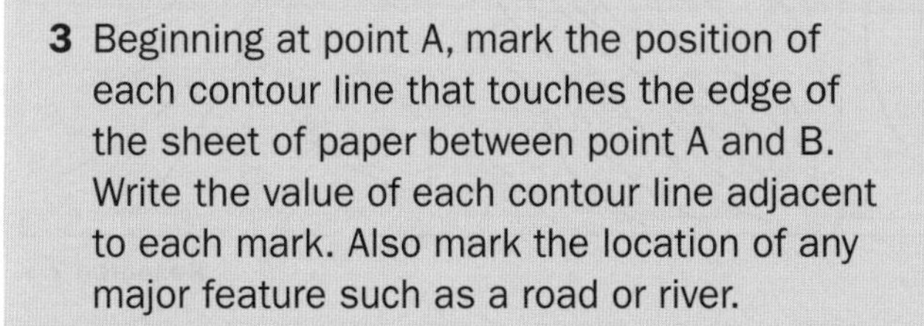

3 Beginning at point A, mark the position of each contour line that touches the edge of the sheet of paper between point A and B. Write the value of each contour line adjacent to each mark. Also mark the location of any major feature such as a road or river.

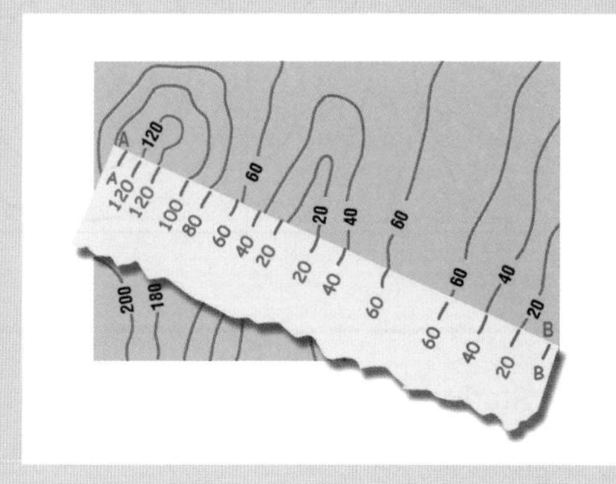

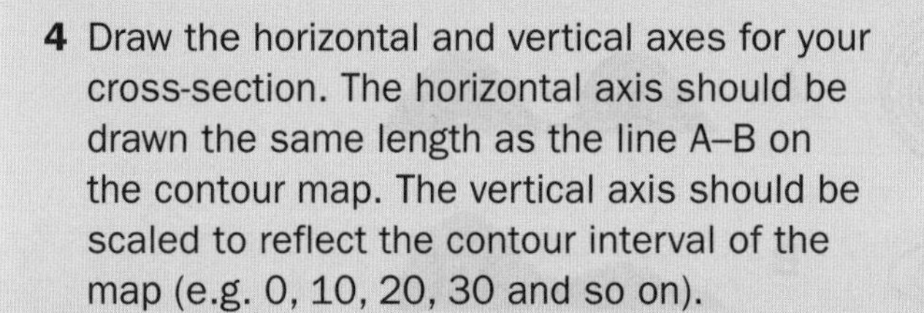

4 Draw the horizontal and vertical axes for your cross-section. The horizontal axis should be drawn the same length as the line A–B on the contour map. The vertical axis should be scaled to reflect the contour interval of the map (e.g. 0, 10, 20, 30 and so on).

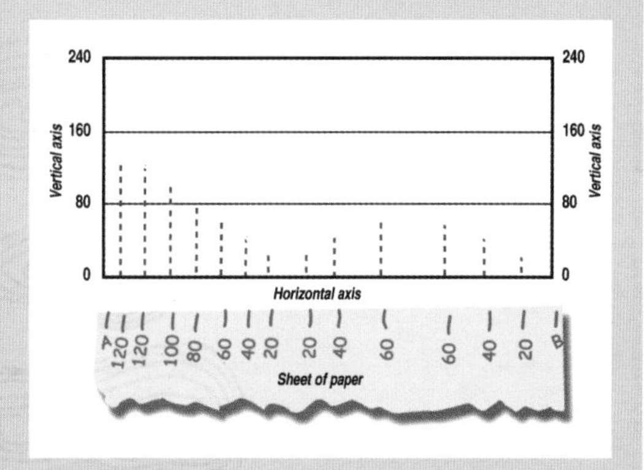

5 Place the sheet of paper along the horizontal axis and plot the contour elevations as though you were plotting a line graph.

Join the dots with a smooth curved line. Draw arrows to show the location of any important features.

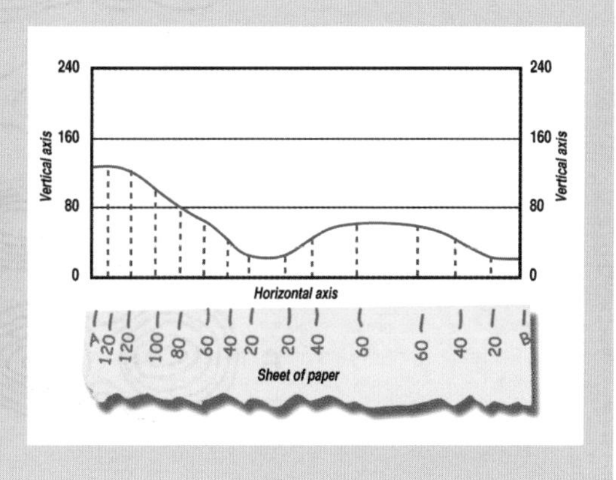

ISBN: 9780170368155

Common contour patterns

A skilled geographer can visualise the shape of the landforms by studying patterns created by contour lines.

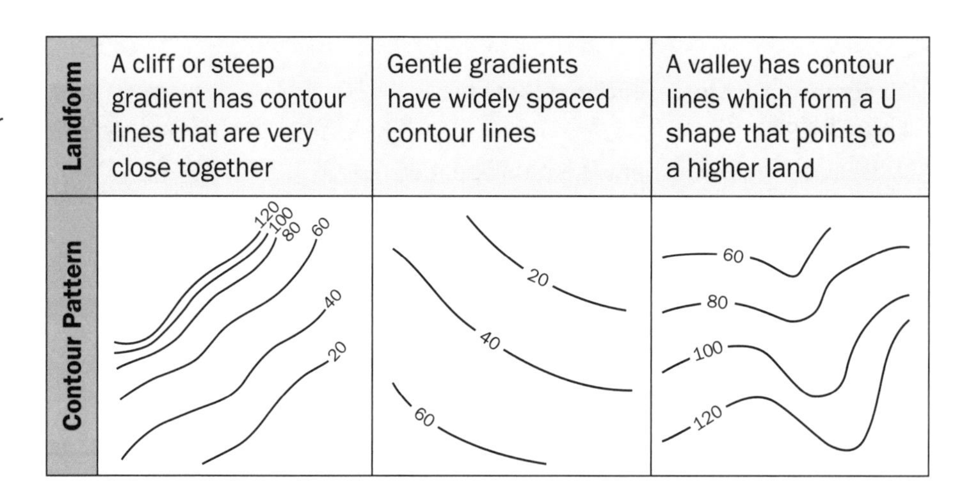

Landform	A cliff or steep gradient has contour lines that are very close together	Gentle gradients have widely spaced contour lines	A valley has contour lines which form a U shape that points to a higher land
Contour Pattern	120, 100, 80, 60, 40, 20	20, 40, 60	60, 80, 100, 120

Landform	A river valley has contour lines which form a V shape that points to higher land	Round hills are shown with enclosed circular contour lines	Enclosed contour lines that have a linear shape indicate a mountain ridge	Lines jutting away from a hill or mountain indicate a spur
Contour Pattern	160, 140, 120, 100	180, 80, 220	80, 100, 120, 120, 100, 80, 60	80, 100, 120, 140, 160

Resource 7.3

Learning Activities

1 Study the patterns in Resource 7.4. Draw a line connecting the contour pattern on the left with the correct landform shape on the right.

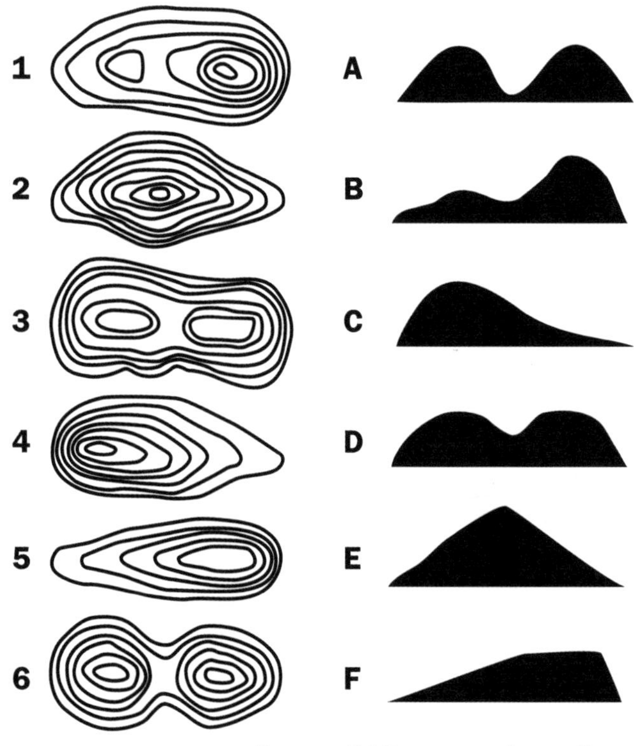

Resource 7.4 Common contour patterns

ISBN: 9780170368155

2 A skilled geographer can visualise the shape of landforms by studying patterns created by contour lines. Using Resource 7.5, construct a cross-section of Rangitoto Island from point A to B using a vertical contour interval of 20 m.

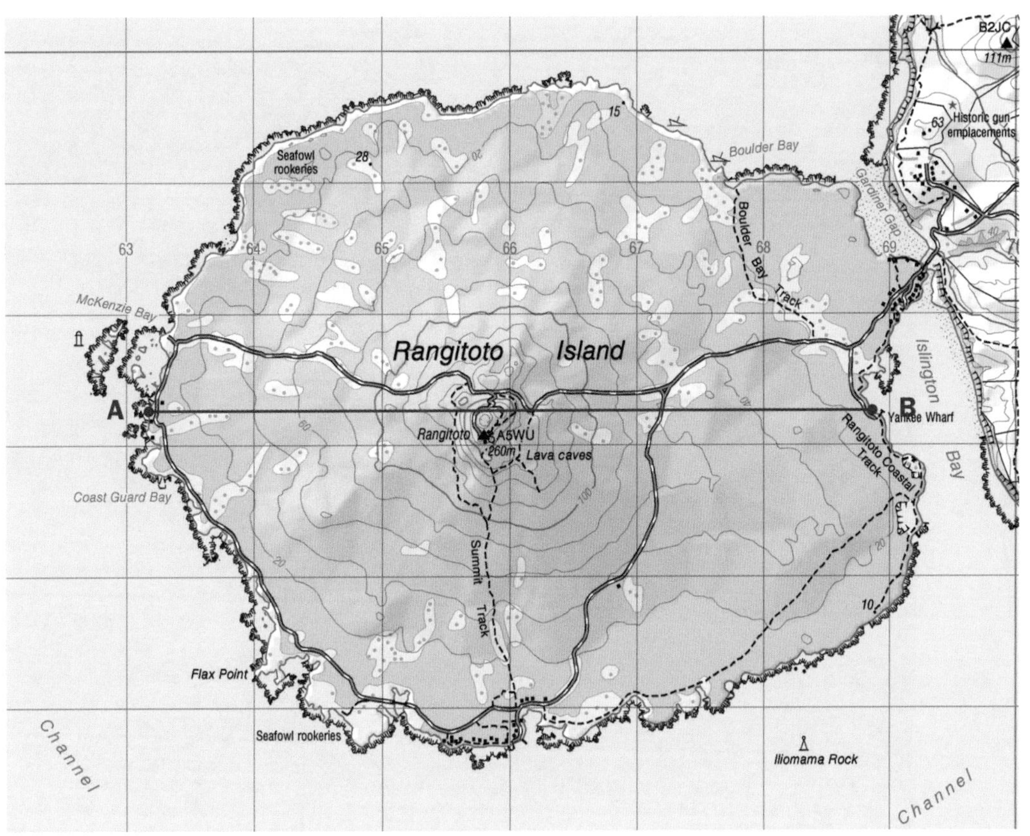

Resource 7.5

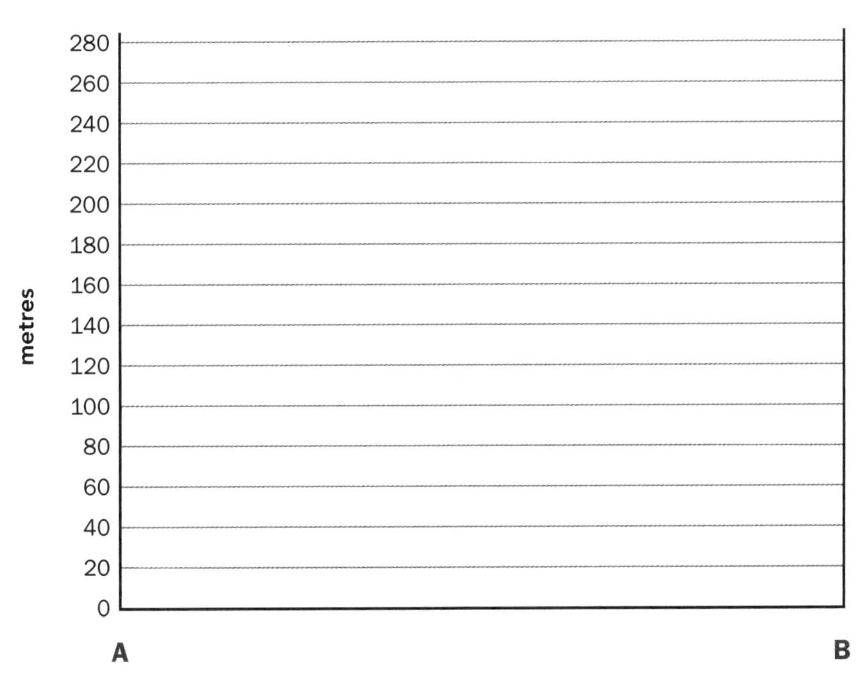

ISBN: 9780170368155

Learning Activities

3 Using Resource 7.6, construct a cross-section of Solander Island from point A to B using a vertical contour interval of 20 m.

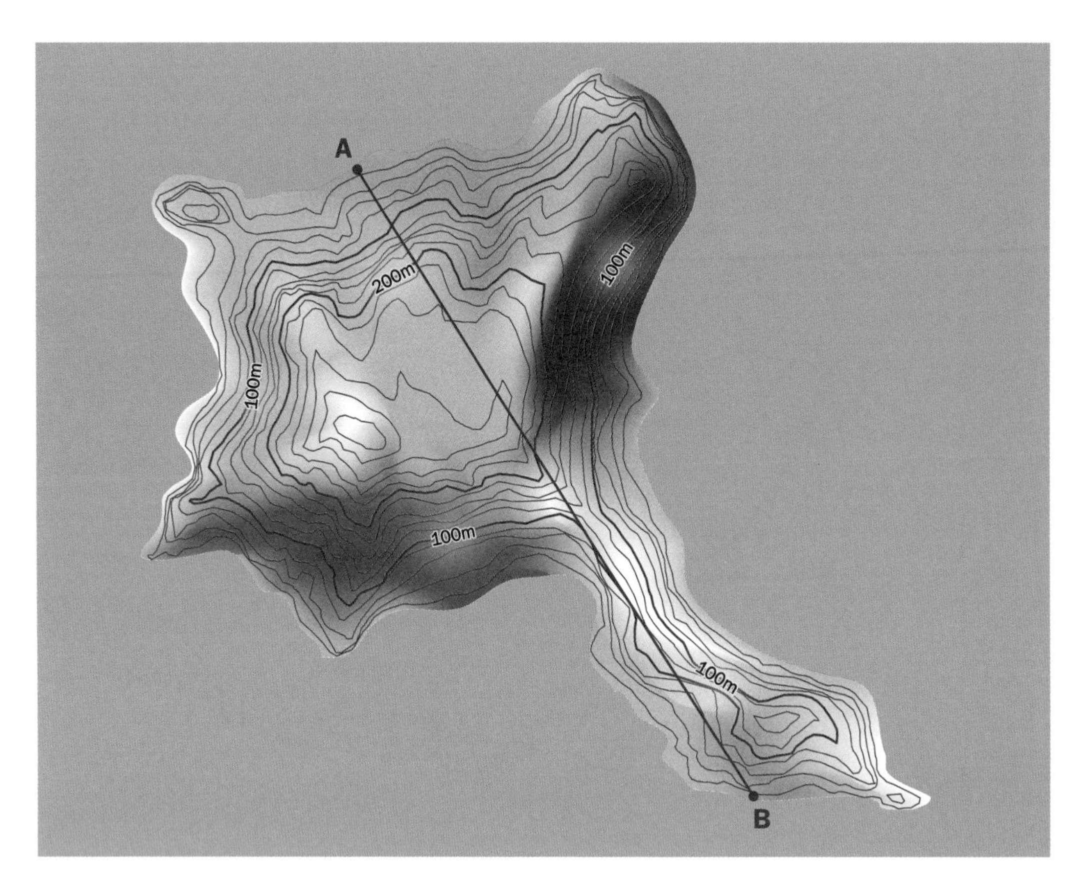

Resource 7.6

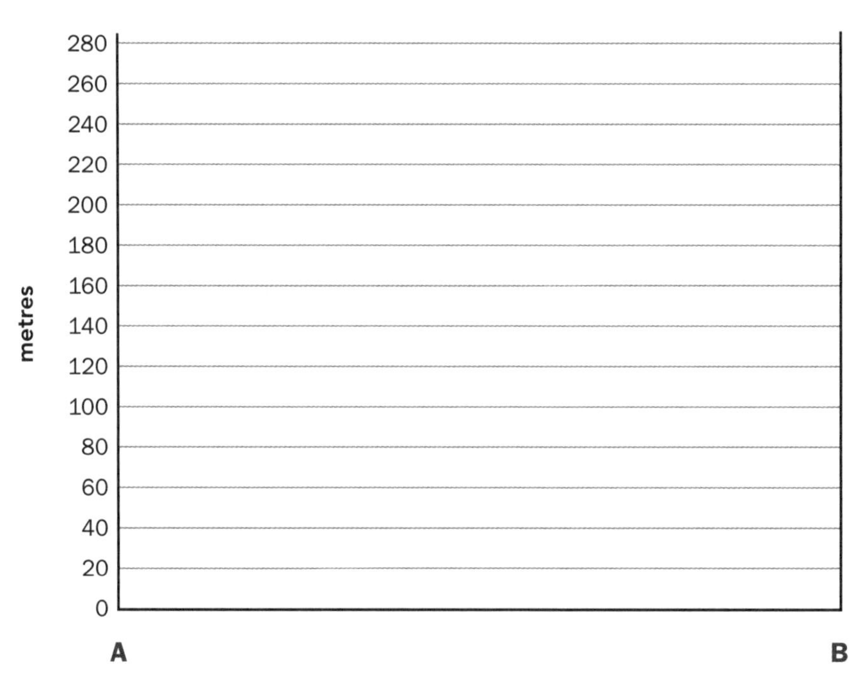

ISBN: 9780170368155

8 Précis map construction

Topographic maps contain an immense amount of detail, so it is sometimes helpful to construct a précis map that highlights just a few of the features illustrated on the map. Précis maps are a useful tool for geographers as they enable the relationship between features to be easily identified.

To construct a précis map, follow the steps below:

1. Identify from the topographic map the particular feature or features you wish to study.
2. Establish the relative location of the feature on the topographic map (or absolute location if a six-figure grid reference is required).
3. Draw a simple outline of each feature on your précis map, taking care to place features in their correct location relative to others. You may choose to draw a grid on your précis that corresponds to the grid on the topographic map to ensure even greater accuracy. Shade each feature in colour pencil using an appropriate colour, for example blue for water features, black for cultural features and green for vegetation.
4. Summarise important features by constructing a key or legend to show the meaning of any symbols or shading used.

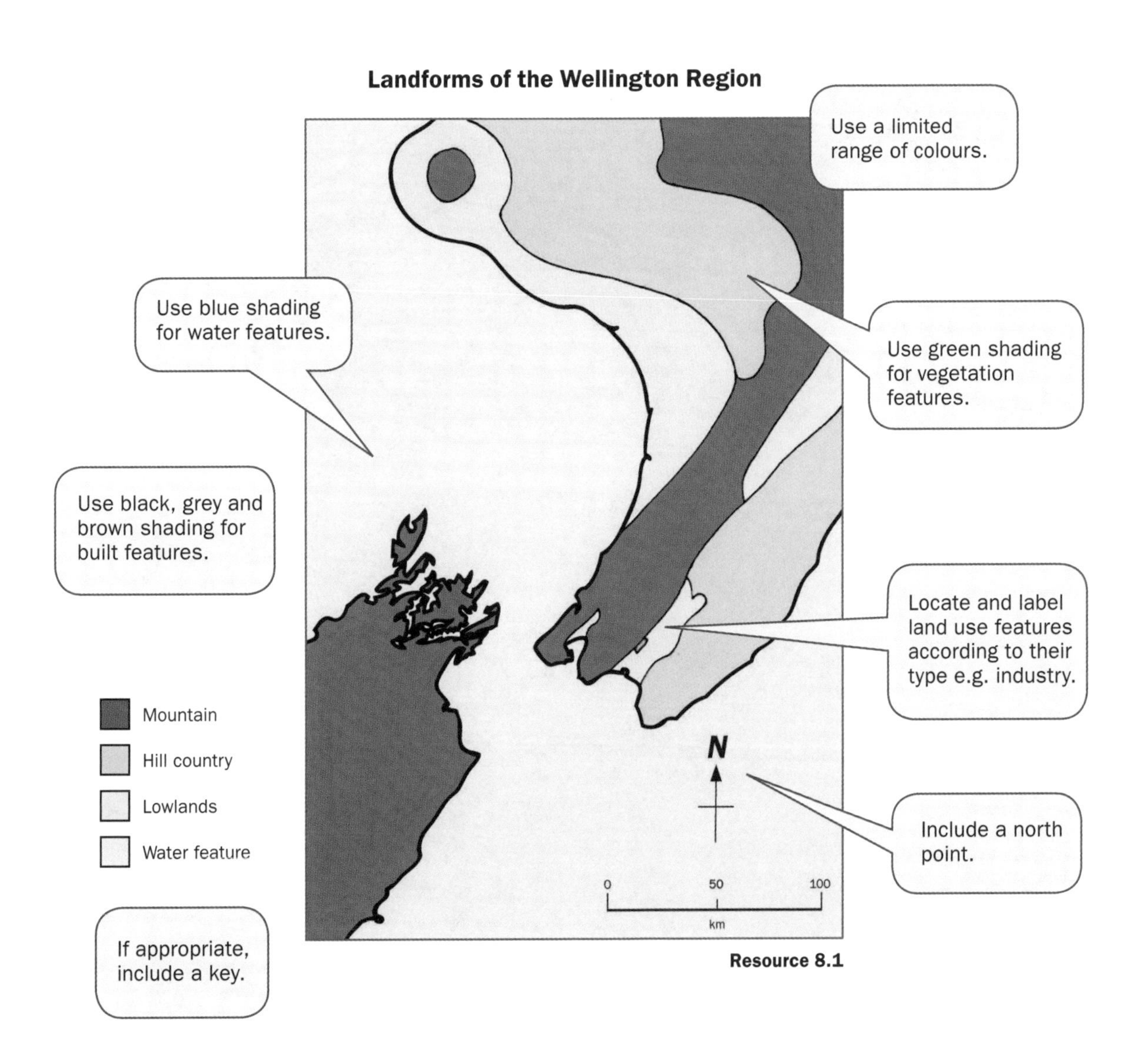

Resource 8.1

ISBN: 9780170368155

Learning Activities

Study the topographic map of Turangi (Resource 8.2).

1 Construct a précis map on the page opposite, which includes the following features:

a Lake Rotoaira

b Pihanga Peak

c State Highway 1 and 46

d Turangi.

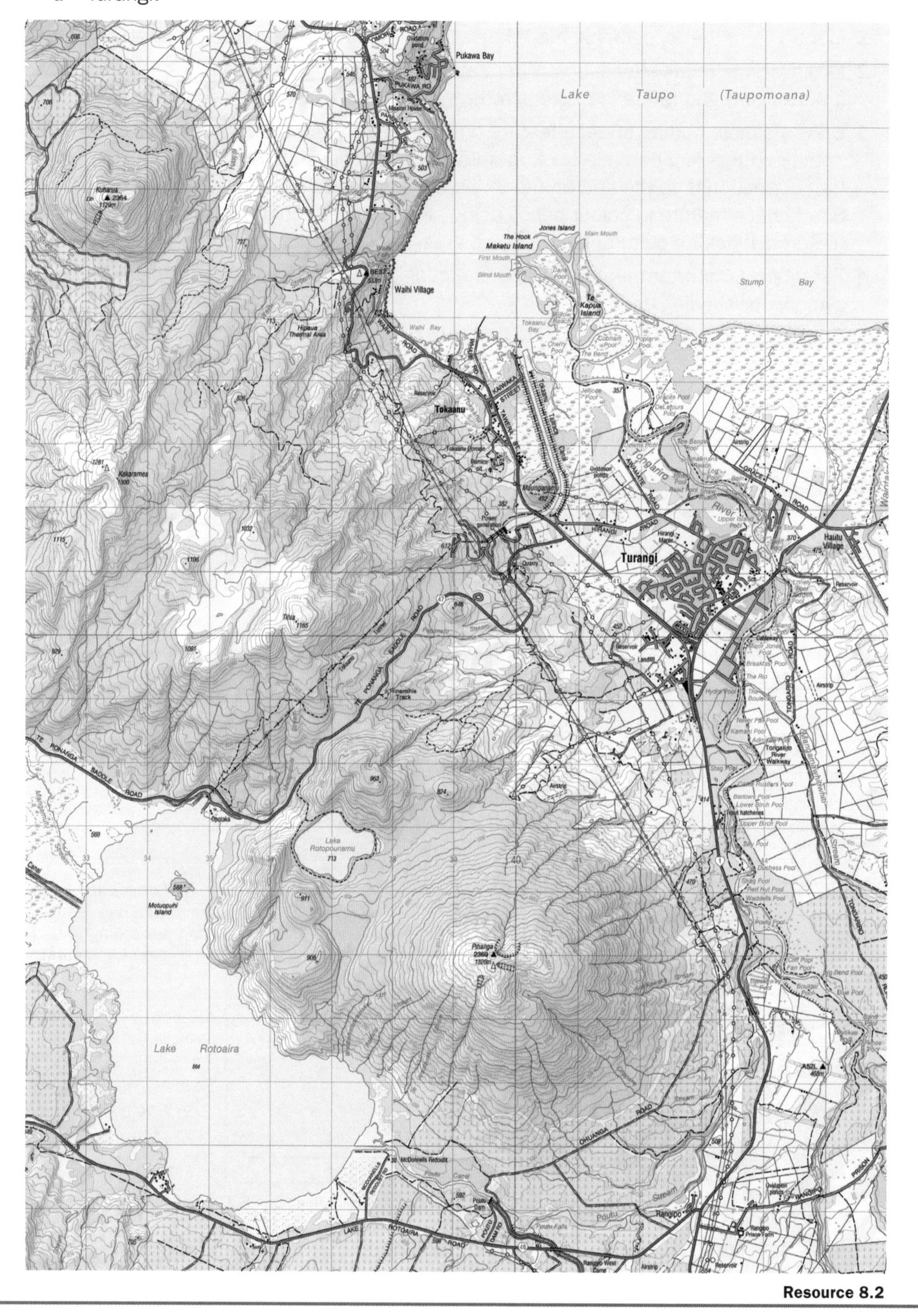

Resource 8.2

ISBN: 9780170368155

Learning Activities

Key

- [ ] Lake Rotoaira
- [ ] Pihanga Peak
- [ ] State Highway 1 and 46
- [ ] Turangi

ISBN: 9780170368155

Learning Activities

Study the topographic map of Hamilton (Resource 8.3).

2 Construct a précis map on the page opposite, which includes the following features:

- **a** Lake Rotoroa
- **b** The urban area east of the Waikato River
- **c** The cultural feature at GR 975177
- **d** The Waikato River.

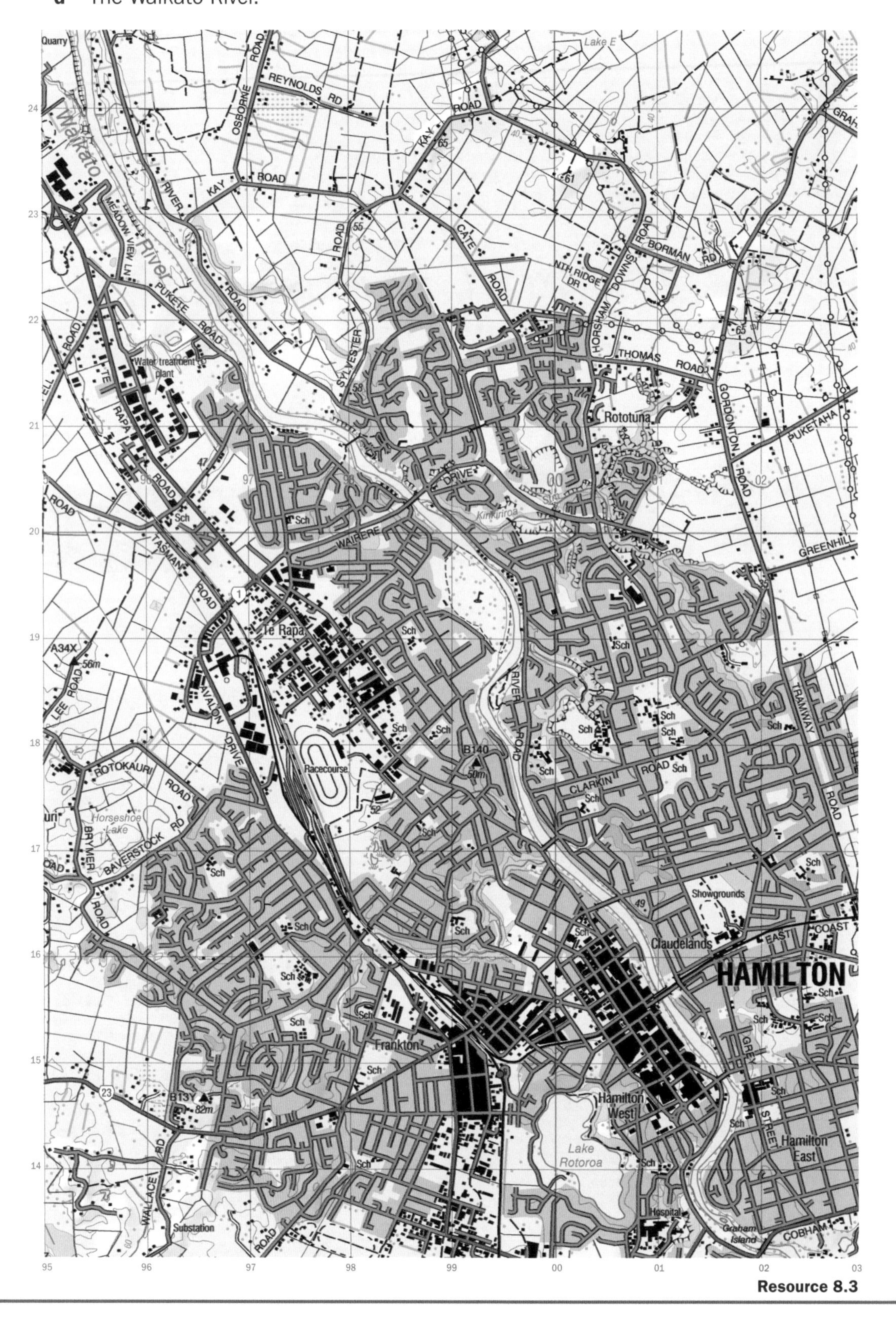

Resource 8.3

ISBN: 9780170368155

Learning Activities

Key

- [ ] Lake Rotoroa
- [ ] The urban area east of the Waikato River
- [ ] The cultural feature at GR 975177
- [ ] The Waikato River

ISBN: 9780170368155

9 Other types of maps

Geographic maps vary in style according to their use and purpose, however all maps can be grouped into one of three categories:

- General maps are constructed for a general audience and contain a variety of information (Resource 9.1). General maps are usually produced as a series, for example New Zealand's 1:250 000 Topo250 topographic map series.
- Thematic maps have specific geographic themes and are oriented towards particular audiences. Examples include dot maps to illustrate population distribution (Resource 9.2), or choropleth maps to illustrate population density (Resource 9.3).
- Topological maps are similar to general maps but are more simplified. They often ignore detail and scale in order to communicate ideas clearly and effectively. The most famous example of a topological map is Beck's London Underground Map.

As an NCEA geographer it is important that you are able to read and interpret the following map types:

Generalised maps

Generalised maps follow all mapping conventions (i.e. scale, north point, key) and contain a variety of information, making them suitable for multiple uses.

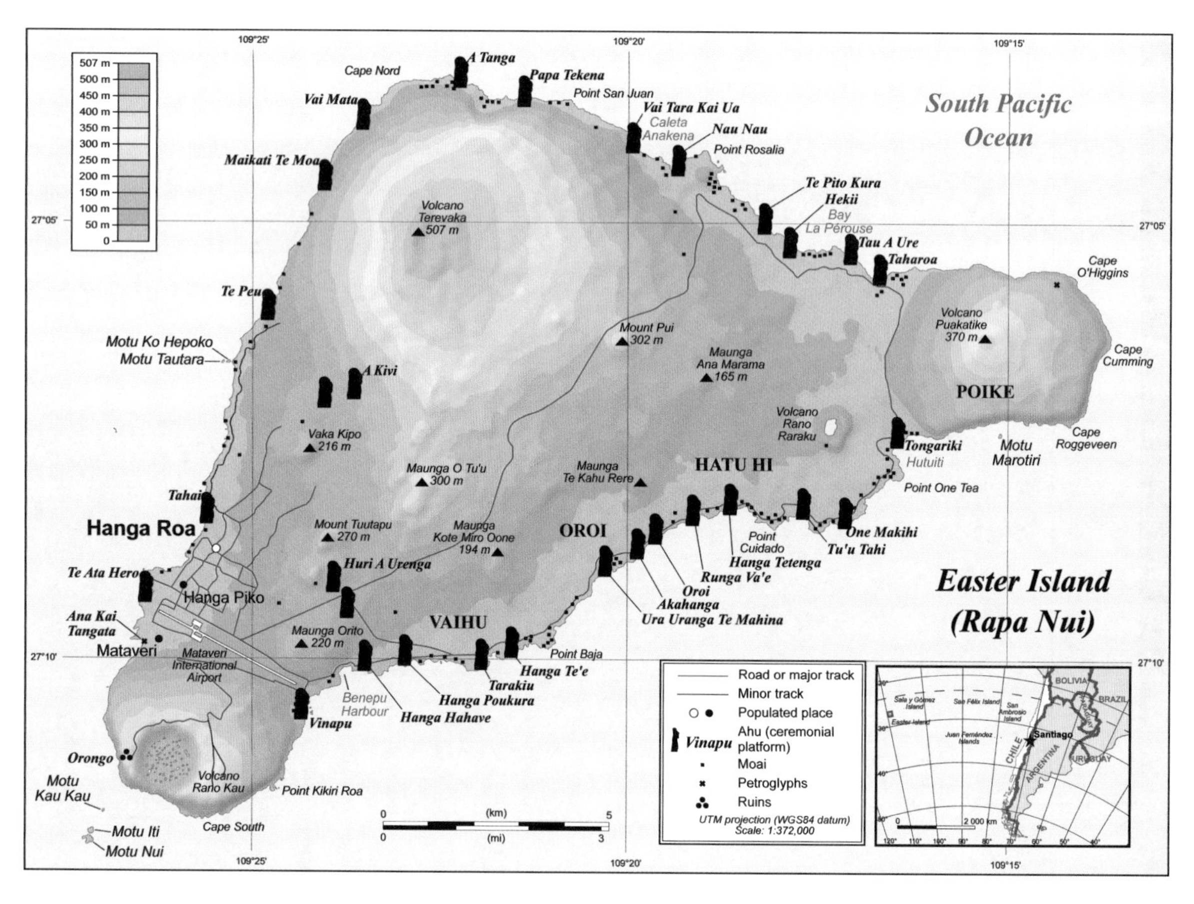

Resource 9.1 Generalised map of Easter Island

ISBN: 9780170368155

Dot distribution maps

Dot distribution maps use a dot symbol to show the presence of a feature, relying on a visual scatter to show spatial pattern.

Resource 9.2 John Snow's map of Soho, showing the distribution of cholera deaths surrounding the Broad Street pump

Choropleth maps

Choropleth maps use different colour shades to show the amount or value of something. Darker shades generally indicate higher concentrations and lighter shades indicate lower concentrations.

Resource 9.3 Choropleth map showing population density in New Zealand

ISBN: 9780170368155

Isoline maps

Isoline maps are made up of lines that join points of equal value. Both contour maps and synoptic (weather) charts fall into this category (Resource 9.4).

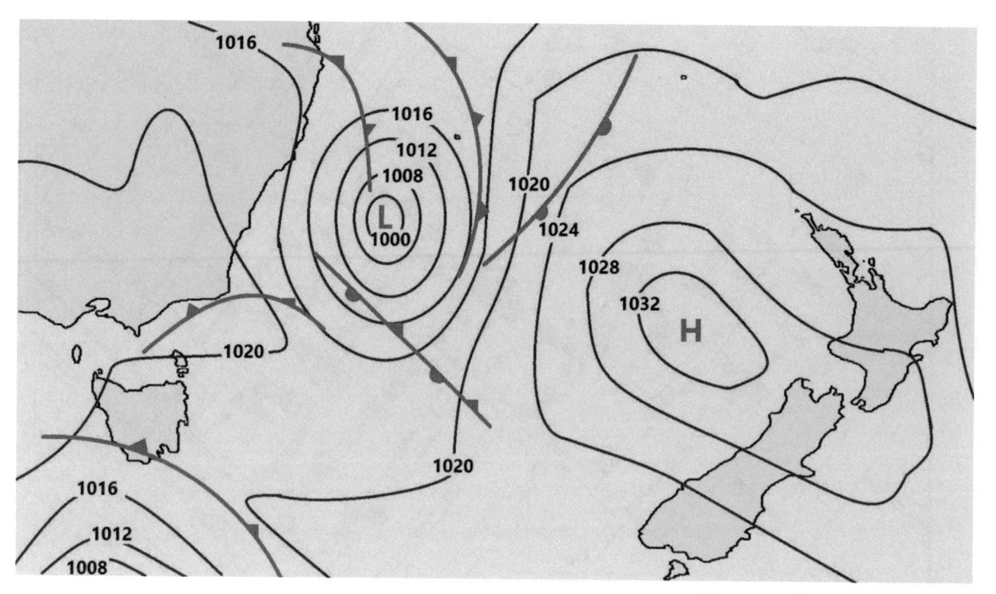

Resource 9.4 Synoptic chart

Flow maps

Flow maps show the movement of objects from one location to another, such as the number of people in a migration, the amount of goods traded between regions or the amount of connections between users of social networks (Resource 9.5). Flow maps sometimes proportionally vary widths of lines to signal the strength or size of a flow or connection.

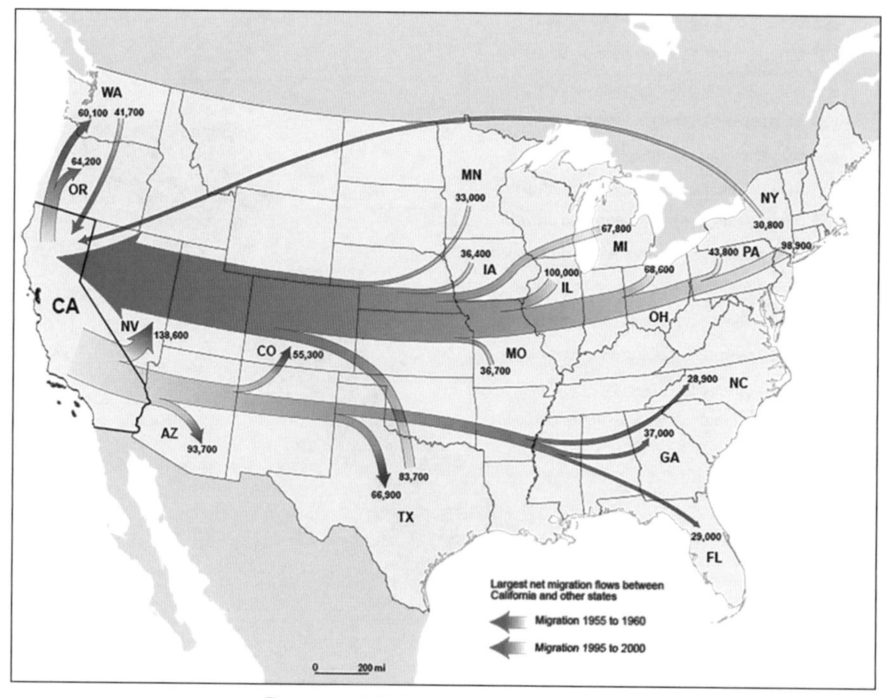

Resource 9.5 Net migration between California and Other States

ISBN: 9780170368155

Cartograms

A cartogram is a type of thematic map in which the areas of spatial features are distorted in proportion to the value of a statistical variable (Resource 9.6).

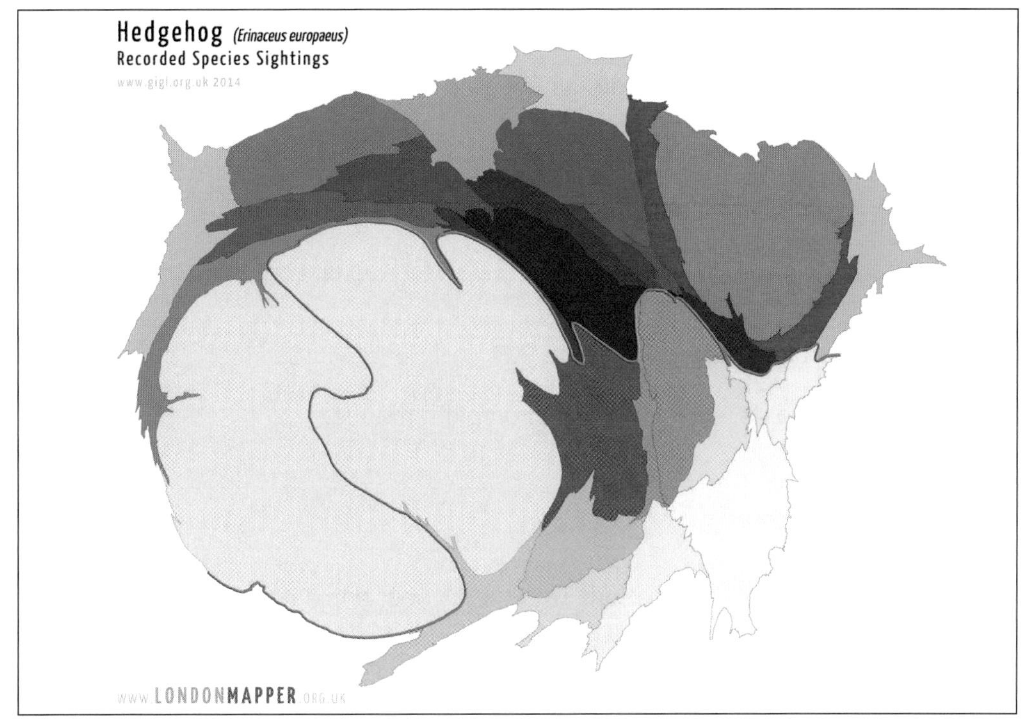

Resource 9.6 The location of London's hedgehog sightings

Topological maps

Topological maps are simplified maps that do not always follow mapping conventions. They usually serve only one purpose (Resource 9.7).

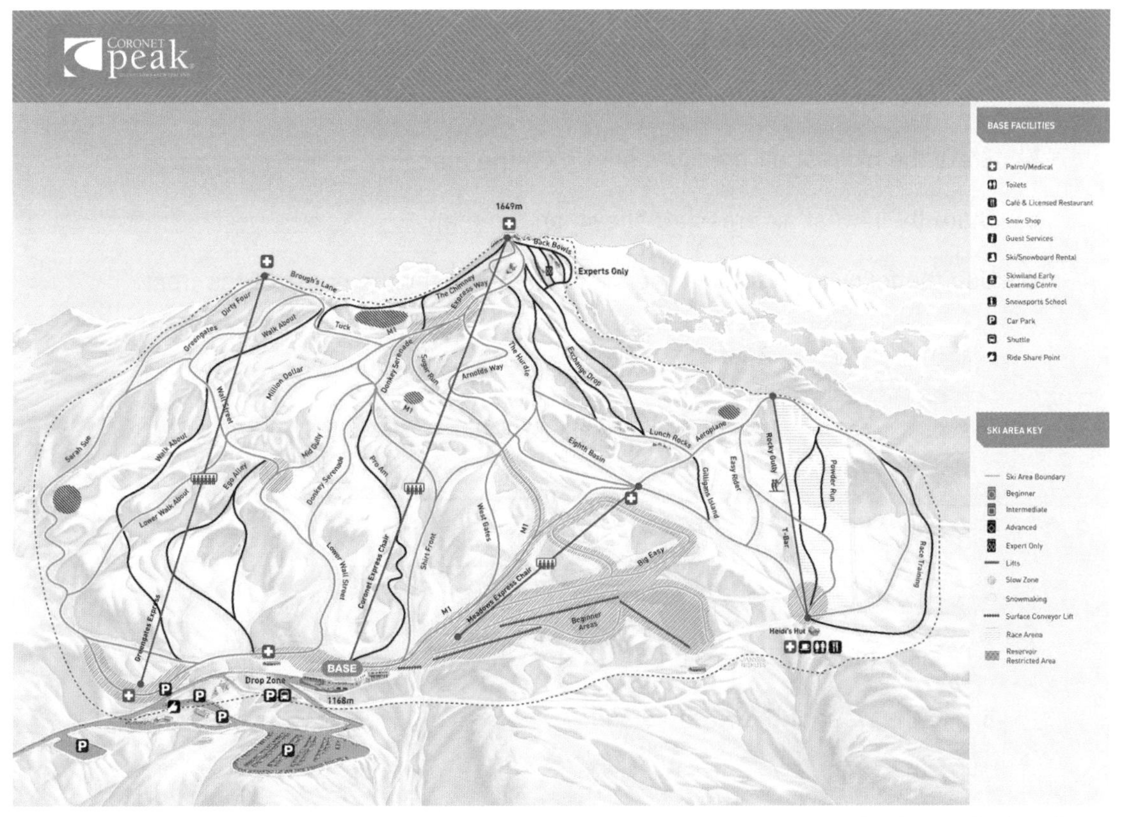

Resource 9.7 Coronet Peak trail map

ISBN: 9780170368155

Learning Activities

1 Study the dot map in Resource 9.2.

a What is the main purpose of a dot map?

b What does the location of dots on the map show?

c Describe the distribution of cholera deaths shown on the map.

2 Study the choropleth map in Resource 9.3.

a In your own words, explain what a choropleth map is.

b State the name of the locations with the highest population density (dark red).

c Name the areas with low population densities (dark green).

3 Study the weather map in Resource 9.4.

a Describe what the isobar lines on the map indicate.

b State the highest air pressure shown on the map.

c State the lowest air pressure shown on the map.

d Use the following weather map symbols to identify the name of the front:

Cold front

Warm front

Stationary front

Occluded front

i South of the low pressure (L) system.

ii To the west of the North Island.

ISBN: 9780170368155

Learning Activities

4 Study the flow line map in Resource 9.5.

a What do the flow lines represent? ____________________

b From which states do most migrants originate?

c Describe the pattern of migration between 1955 and 1960.

d Describe the pattern of migration between 1995 and 2000.

5 Study the cartogram in Resource 9.6

a State the name of UK city represented in the map. ____________________

b Where are most hedgehogs reportedly found?

6 Study Resource 9.7 and complete the following questions.

a State the main purpose of the map. ____________________

b Identify three ways the Coronet Peak route map does not follow normal mapping conventions.

c What kind of ski trails are:

i concentrated near the summit? ____________________

ii adjacent to Meadows Express Chair? ____________________

iii most widespread? ____________________

d What ability are the skiers who dine at Heidi's Hut likely to be? ____________________

ISBN: 9780170368155

Graph interpretation and construction

Graphs offer geographers simple and effective ways of presenting statistical information. Geographers regularly employ graphs to:

- Compare two or more variables
- Illustrate change over time
- Illustrate the relationship between two sets of data
- Identify patterns or trends
- Show how something is made up.

There are many types of graphs, each designed to present information in a specific way. Since each graph type is suited for illustrating different types of information, it is important you learn how to interpret and construct a variety of graph types. In this chapter, you will learn how to interpret and construct bar, line, pie and percentage bar graphs, scatter graphs, pictograms, age-sex pyramids and climate graphs.

10 Bar graphs (single and multiple)

Bar graphs are the simplest way to compare two sets of information. They generally use horizontal bars, but column graphs use vertical bars. For the purposes of NCEA Geography, however, you will not usually be required to make a distinction between horizontal bar graphs and vertical column graphs.

To construct a simple bar graph, follow the steps below:

1. Decide what information is to be plotted on each axis. In most cases you will plot the non-quantifiable variable (i.e. the one that does not change) on the *x*-axis (e.g. country names, age groups, or periods such as months or years), while the quantifiable data is normally plotted on the *y*-axis. It is for this reason the *y*-axis is sometimes referred to as the variable axis.
2. Determine an appropriate scale for the variable axis. Bar graphs abide by the graphing convention that requires the *y*-axis (variable axis) to follow a constant number scale starting from zero (e.g. 0, 5, 10, 15 or 0, 10, 20, 30).
3. Having determined the range and scale of the data to be plotted, use a ruler to construct the axes. Like most graphs, bar graphs have two axes: the *x*-axis is usually horizontal (i.e. runs across the bottom of the graph), while the *y*-axis is usually vertical (i.e. runs up the left side of the graph). Ensure both axes are long enough to accommodate the range of data you wish to show.
4. Label each axes (including the units of measurement) and give the graph a title that clearly states what it illustrates. If appropriate, the title should also include location and date specific information.
5. Use a ruler to construct each bar. Ensure all bars are drawn of equal width and use consistent spacing.
6. Shade each bar with colour pencil. Label each bar, or else include a key if you are constructing a multiple bar graph (Resource 10.2).

ISBN: 9780170368155

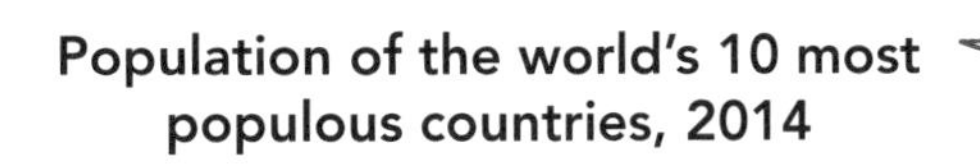

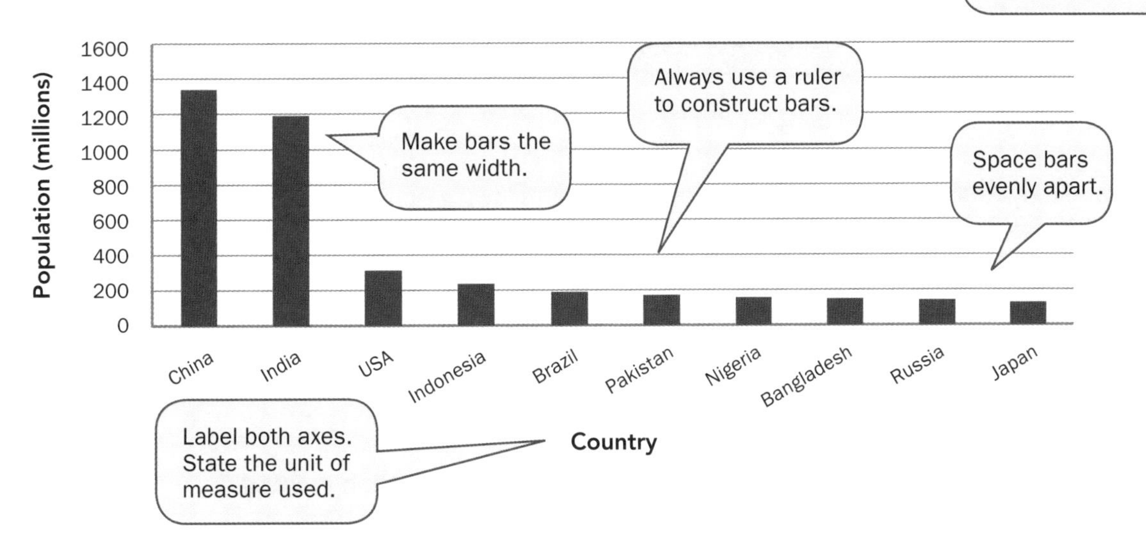

Resource 10.1 Simple bar graph showing the world's 10 most populous countries (2014)

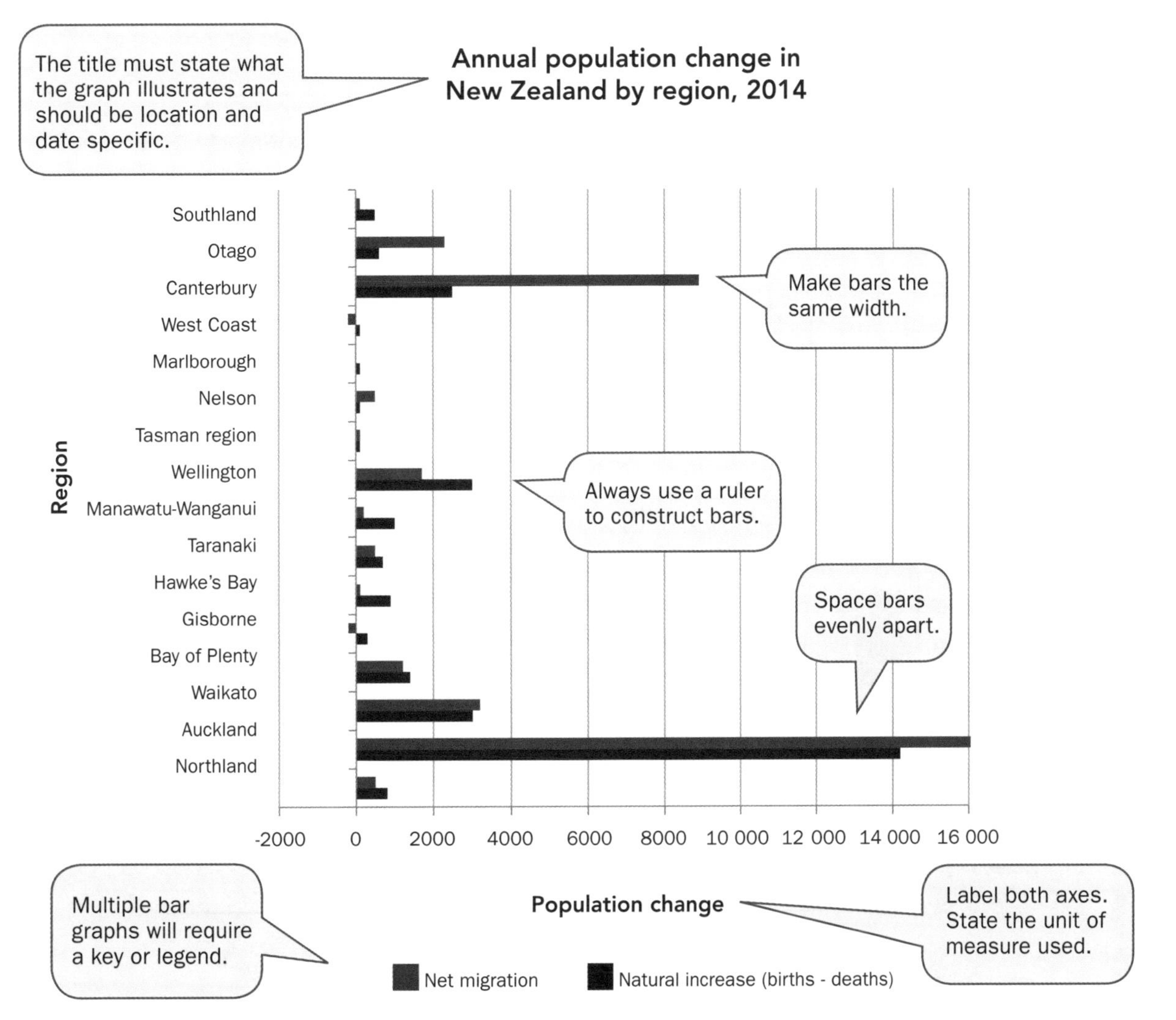

Resource 10.2 Multiple bar graph showing annual population change in New Zealand by region (2014)

ISBN: 9780170368155

Learning Activities

1 Study Resource 10.1 and then complete the following activities.

a State the population of China. ______

b Estimate the number by which the population of India exceeded that of USA in 2014.

______

2 Study Resource 10.2 and then complete the following activities.

a State the natural increase in population for Wellington in 2014. ______

b Name the region of New Zealand that experienced the largest population increase in 2014.

______

c Name the two regions that experienced negative migration in 2014.

______

d Which component of population change (i.e. net migration or natural increase) had the greatest impact on the population in 2014?

______

3 Use the data in Table 10.1 to construct a bar graph showing infant mortality rates in selected countries.

Afghanistan	Nigeria	Pakistan	India	Iraq	Samoa	USA	New Zealand	Singapore
152	93	65	49	43	23	6	5	2

Table 10.1 Infant mortality rates (per 1000) for selected countries

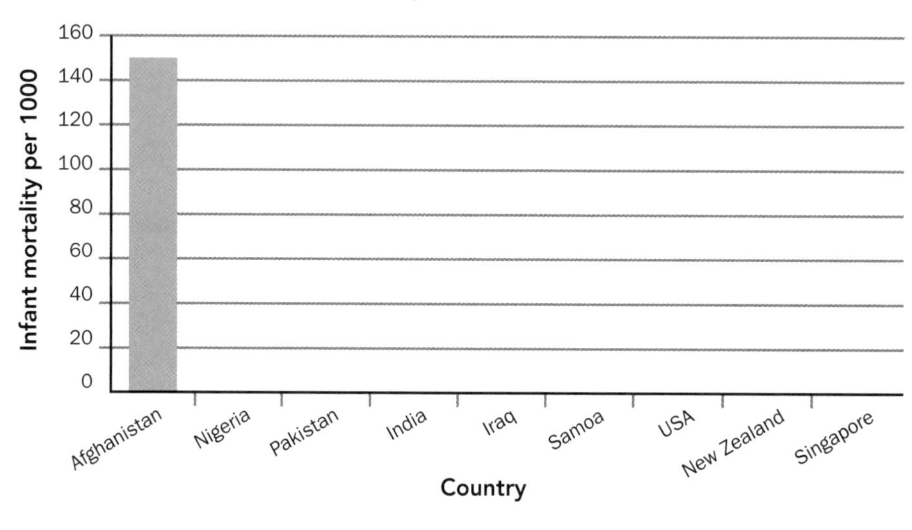

ISBN: 9780170368155

4 Use the data in Table 10.2 to construct a bar graph showing the costliest natural hazards for insurers this century.

Hazard Event	Hurricane Katrina, USA (2005)	Japan Quake & Tsunami (2011)	Hurricane Sandy, USA (2012)	Hurricane Ike, USA (2008)	Christchurch Quake (2011)	Chilean Quake (2010)
US$ (billions)	71	35	29	20	12	9

Table 10.2 Costliest natural hazards this century

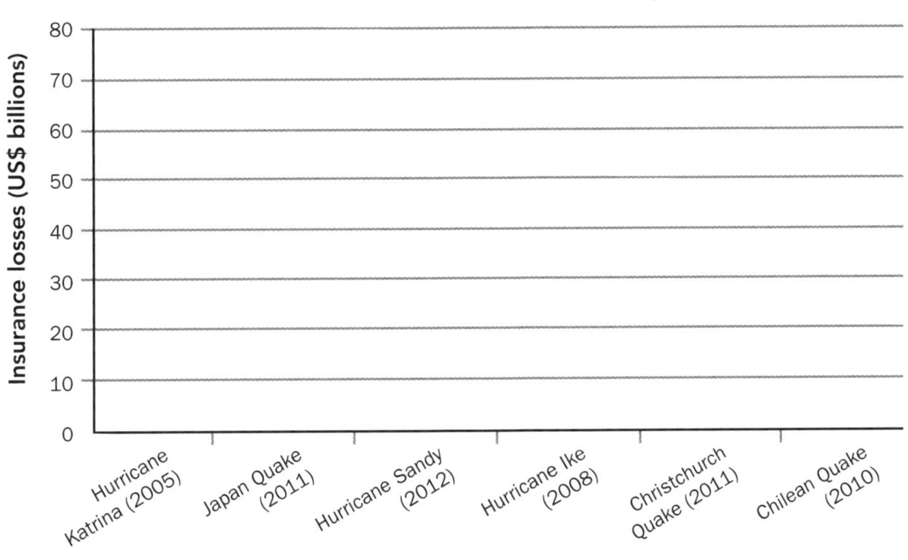

5 Use the data in Table 10.3 to construct a multiple bar graph showing the percentage of government expenditure on education and the military for selected countries.

	USA	United Kingdom	Australia	New Zealand	Russia	India	China
Education	5.5	5.6	4.7	6.2	3.9	3.2	1.9
Military	4.6	2.4	3.0	1.0	3.9	2.5	4.3

Table 10.3 Percentage of government expenditure on education and the military for selected countries

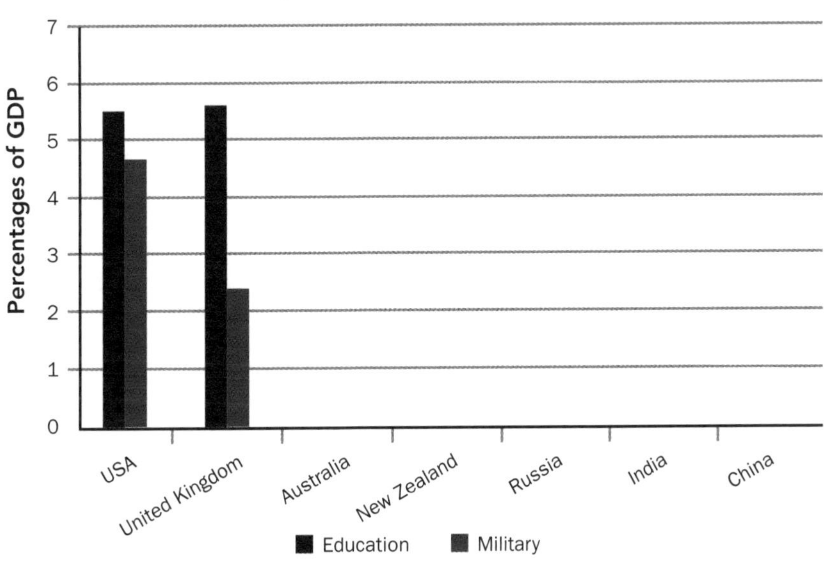

ISBN: 9780170368155

11 Line graphs (single and multiple)

Line graphs are especially useful in geography as they are easy to create, while their visual characteristics reveal data trends clearly.

Like bar graphs, line graphs provide a visual representation (shown on the *x*-axis and *y*-axis) of how two variables relate. In a line graph the vertical *y*-axis usually indicates quantity (e.g. population size or change or volume), or percentage in the case of a compound graph. The horizontal *x*-axis often measures units of time (e.g. years). As a result, the line graph is often used to show change over time. If you wanted to graph population change over time, for example, you could measure the time variable in years along the *x*-axis, and population size along the *y*-axis (Resource 11.1).

Multiple line graphs differ from single line graphs in that they allow two or more related sets of data to be plotted on the same graph, allowing for easy data comparison (Resource 11.2).

Not all graphs show information from the past. Some are used to show estimates about the future. These estimates are called *projections* (that is, what we believe will happen in the future, based on our interpretation of past and current trends). For example, the line graph in Resource 11.1 shows that world population is estimated to peak near 10 billion sometime within the next 100 to 200 years. This projection was determined by studying population growth rates in the rich and poor countries of the world.

Projected data is usually shown on a line graph by a dotted or dashed line.

To construct a simple line graph, follow the steps below:

1. Decide what information is to be plotted on each axis. In most cases you will plot the non-quantifiable variable (i.e. the one that does not change) on the *x*-axis (e.g. country names, age groups, or periods such as months or years), while the quantifiable data is normally plotted on the *y*-axis. It is for this reason the *y*-axis is sometimes referred to as the variable axis.
2. Determine an appropriate scale for the variable axis. Like bar graphs, line graphs adhere to the graphing convention that requires axes to follow a constant number scale starting from zero (e.g. 0, 5, 10, 15 or 0, 10, 20, 30).
3. Having determined the range and scale of the data to be plotted, use a ruler to construct the axes. Like most graphs, simple line graphs have two axes: the *x*-axis is usually horizontal (i.e. runs across the bottom of the graph) while the *y*-axis is usually vertical (i.e. runs up the left side of the graph). Ensure both axes are long enough to accommodate the range of data you wish to show.
4. Label each axes (including the units of measurement) and give the graph a title that clearly states what the graph illustrates. If appropriate, the title should also include location and date specific information.
5. Next, plot each value on the graph, and then join the points together either with a ruler (for a straight line curve) or freehand (if a smooth curve is required).
6. If you are constructing a multiple line graph, it is recommended that you also include a key (Resource 11.2).

ISBN: 9780170368155

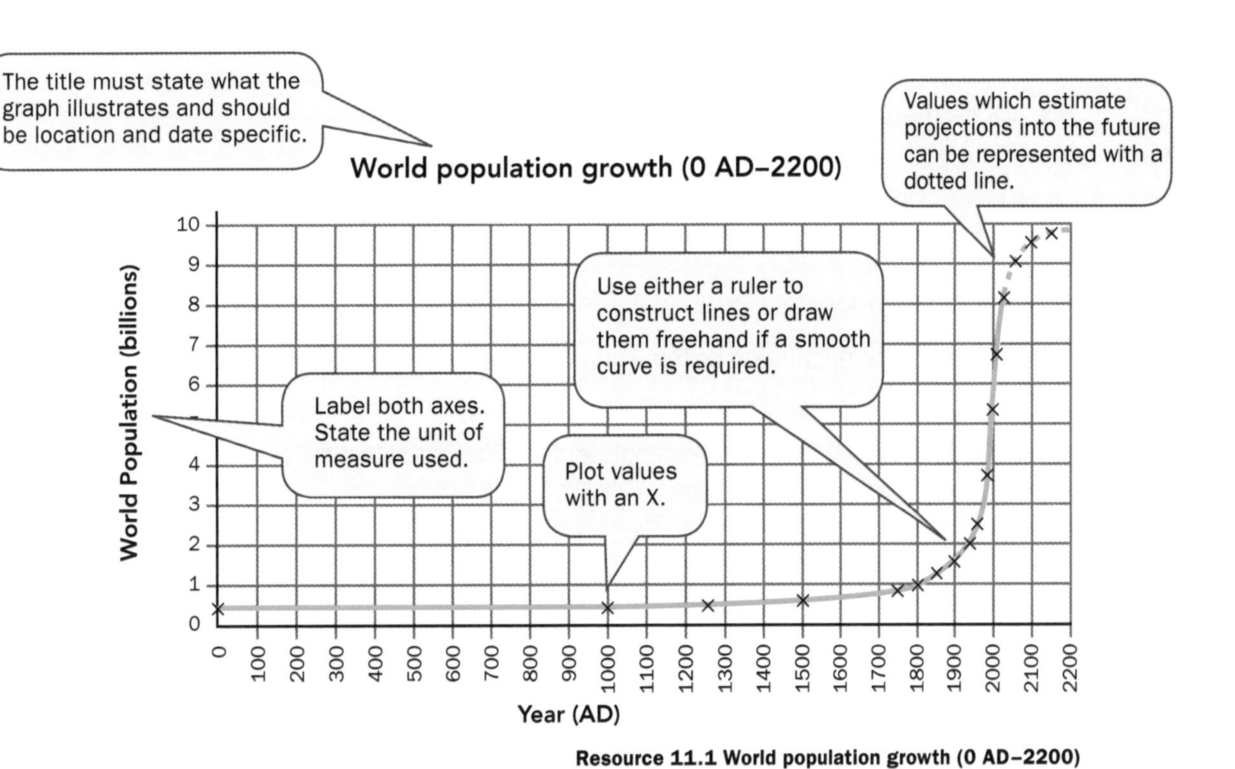

Resource 11.1 World population growth (0 AD–2200)

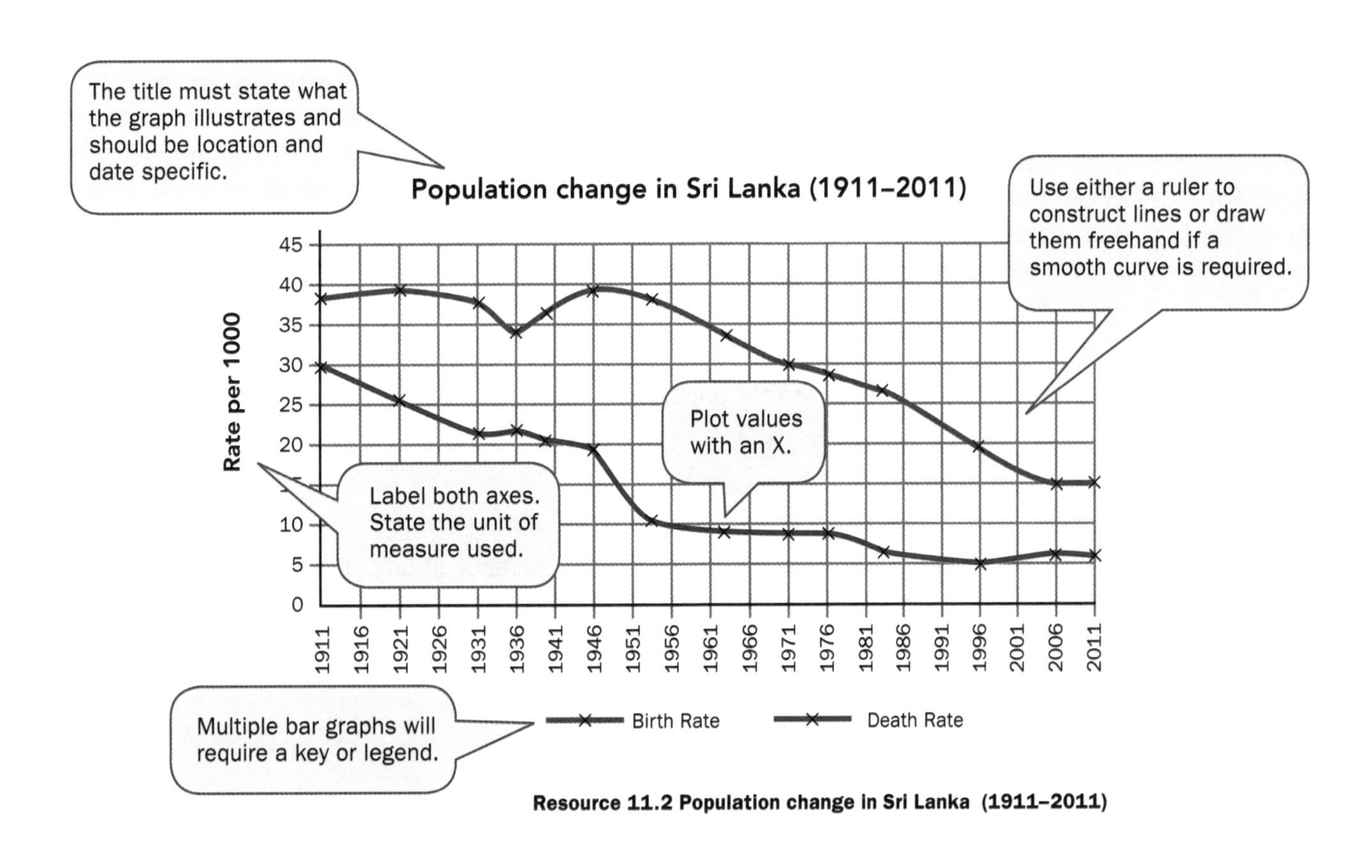

Resource 11.2 Population change in Sri Lanka (1911–2011)

ISBN: 9780170368155

Learning Activities

1 Study Resource 11.1 and then complete the following activities.

a State the population of the world in:

i 1800 ______

ii 1900 ______

iii 2000 ______

b Estimate the population of the world in 2100. ______

2 Study Resource 11.2 and then complete the following activities.

a State the birth rate (per 1000) for Sri Lanka in:

i 1911 ______

ii 1961 ______

iii 2011 ______

b State the death rate (per 1000) for Sri Lanka in:

i 1911 ______

ii 1961 ______

iii 2011 ______

c The rate of natural increase (NI) for a population can be found by calculating the difference between a country's birth rate (BR) and death rate (DR). Using the following equation, calculate the rate of natural population increase for Sri Lanka in 1911, 1961 and 2011:

NI = BR – DR

______

______

______

d Identify the years Sri Lanka's population growth rate was at its:

i fastest ______

ii slowest ______

e In what way has the rate of natural population increase in Sri Lanka changed over time?

______

______

______

______

______

ISBN: 9780170368155

3 Use the data in Table 11.1 to construct a line graph to show population change in New Zealand from 1951 to 2011.

Year	Population (millions)
1951	2.0
1961	2.5
1971	2.9
1981	3.1
1991	3.5
2001	3.9
2011	4.4
2016	4.5*

*** estimated**

Table 11.1 Population change in New Zealand (1951-2011)

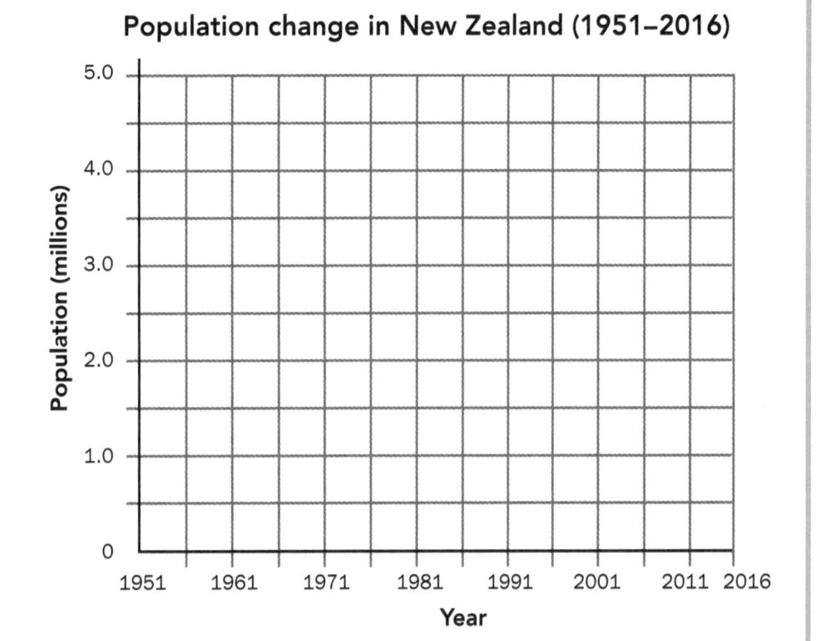

4 Use the data in Table 11.2 to construct a multiple line graph showing population change in selected regions of New Zealand.

Census year	Auckland region	Waikato region	Wellington region	Canterbury region
1991	74 223	10 542	24 129	14 910
1996	107 580	14 151	23 637	25 278
2001	128 019	15 162	25 701	24 606
2006	169 497	23 679	36 588	40 965

Table 11.2 Population change in selected regions of New Zealand

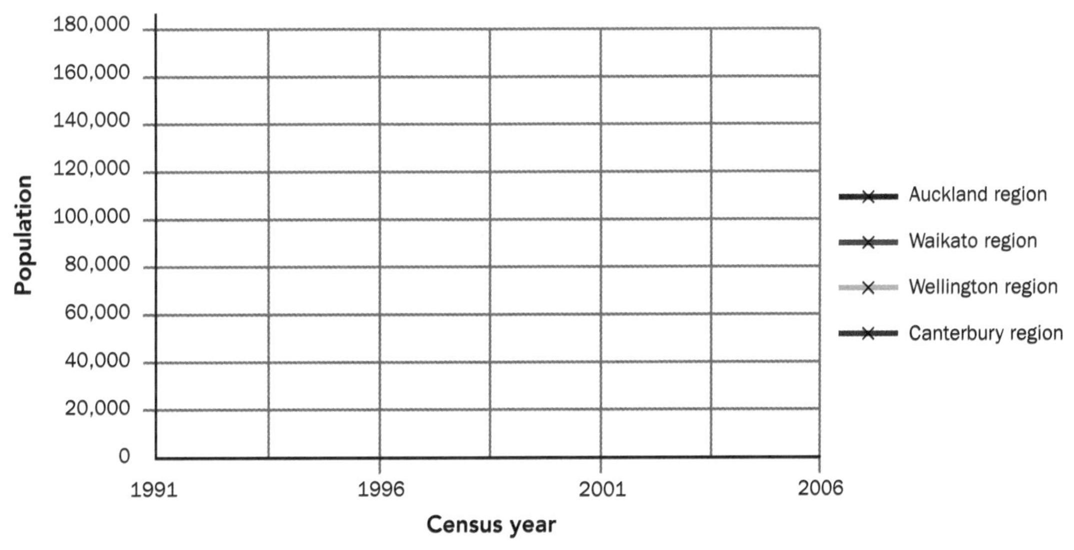

ISBN: 9780170368155

Pie graphs

A pie graph is a useful way of summarising data that has been categorised. A pie graph is circular and is divided into a series of segments. Each segment represents a particular category. The area of each segment is the same proportion of the circle as the category is to the total data set. The pie chart in Resource 12.1, for example, clearly shows that 53% of New Zealand's electricity generation comes from hydropower, and that only 5% comes from wind power.

To construct a pie graph, follow the steps below:

1. Use a compass or stencil to construct a circle. Then draw a line from the centre of the circle to the 12 o'clock position.
2. Convert each percentage value into degrees by multiplying each variable by a factor of 3.6. In Resource 12.1 the hydro portion of the graph was calculated by multiplying the percentage of electricity generated by hydro by 3.6 to represent 201.6˚ (53% x 3.6 = 190.8°).
3. Working clockwise from the 12 o'clock position use a protractor to plot each segment of the pie, beginning with largest category first followed by the second largest category and so on. If applicable, plot the 'Other' category last.
4. Shade each segment with coloured pencils. Label each segment, or alternatively include a key.
5. Give the graph a title that clearly states what it illustrates. The title should also include any location and date specific information.

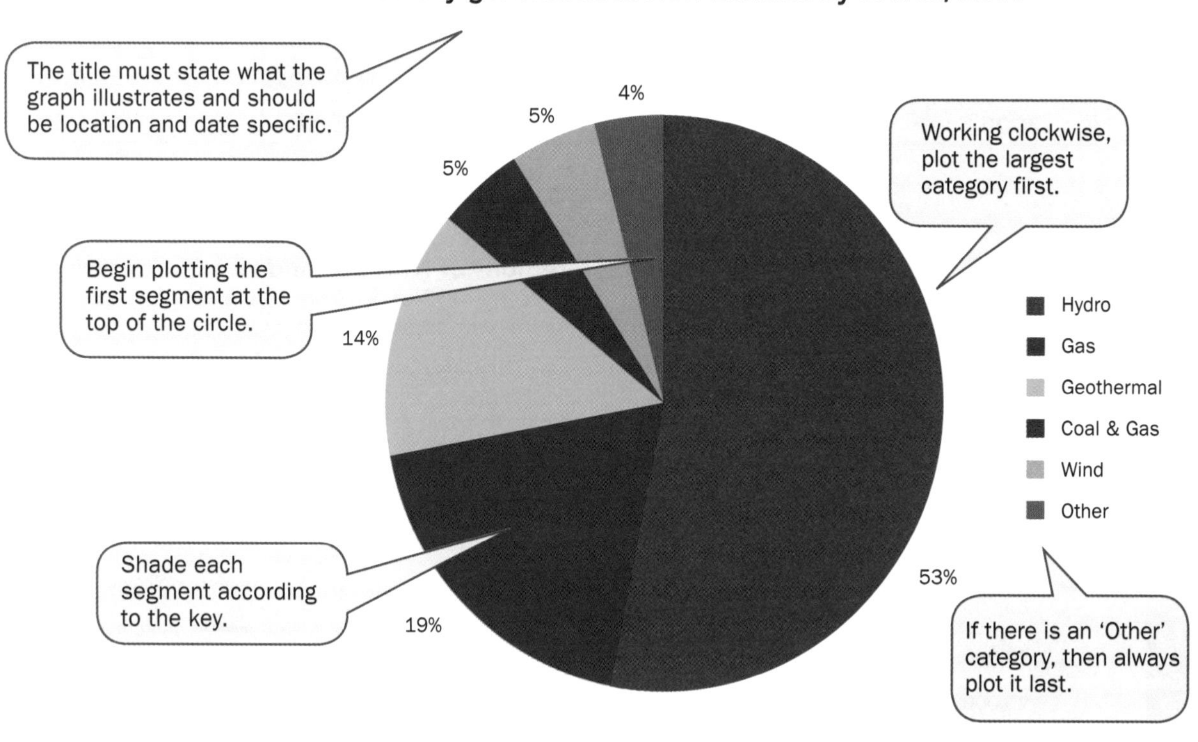

Resource 12.1 Electricity generation in New Zealand by source (2014)

ISBN: 9780170368155

Percentage bar graphs

Like the pie graph, the percentage bar provides a way of summarising data that has been categorised or represents different values of a given variable. A percentage bar graph differs to a pie graph, however, in that it is constructed in the form of a bar rather than a circle (Resource 12.2). To allow for easy conversion from percentage to segment length, percentage bars are usually drawn 100 mm long.

To construct a percentage bar graph, follow the steps below:

1. Use a ruler to construct a horizontal bar measuring 100 mm by 10 mm.
2. Convert each percentage variable into an equivalent proportional distance. For instance a variable of 25% would be equal to a segment length of 25 mm, and a variable of 14% would be equal to a segment length of 14 mm.
3. Working from the left end of the bar, measure and plot each segment beginning with largest category, followed by the second largest category and so on. If applicable, plot the 'Other' category last.
4. Shade each segment with coloured pencils. Label each segment, or alternatively include a key.
5. Give the graph a title that clearly states what the graph illustrates. The title should also include location and date specific information.

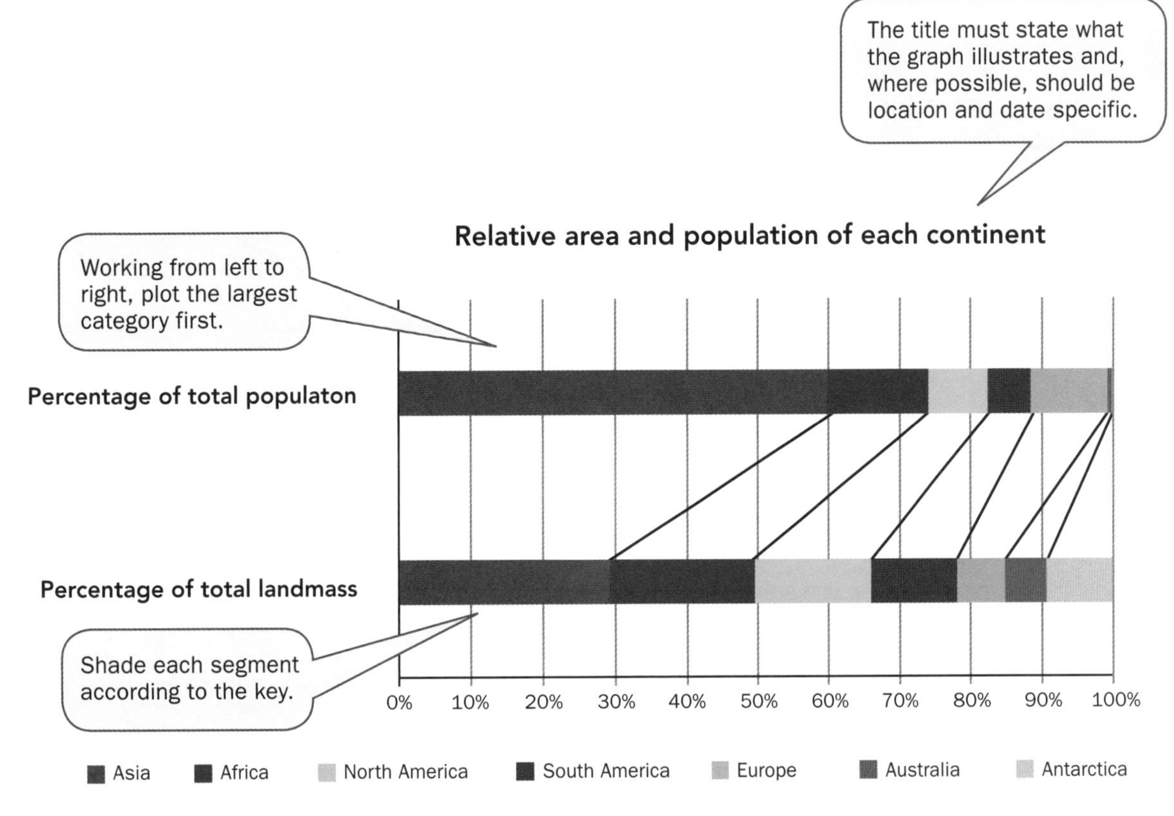

Resource 12.2 Relative area and population of each continent

ISBN: 9780170368155

Learning Activities

1 Study Resource 12.1 and then complete the following activities.

a State the main source of electricity generation in New Zealand. ______________

b List the sources of electricity generated from:

i Non-renewable energy sources.

____

ii Renewable energy sources.

____

c Calculate the percentage of electricity generated from:

i Non-renewable energy sources.

____

ii Renewable energy sources.

____

2 Use the data in Table 12.1 to construct a pie graph showing the ethnic composition of the New Zealand's population in 2011.

Ethnicity	%
European	68
Māori	14
Asian	11
Pacific	6
Other	1

Table 12.1 Ethnic population of New Zealand (2013)

Ethnic population of New Zealand (2013)

☐ European ☐ Māori ☐ Asian

☐ Pacific ☐ Other

ISBN: 9780170368155

Learning Activities

3 Use the data in Table 12.2 to construct a pie graph showing social network users by age. (Hint: Calculate the percentage of social network users for each age group first.)

Age group (years)	Social network users (millions)	%
18–24	28.3	
25–34	35.3	
35–44	29.7	
45–54	26.4	
55–64	19.7	
65+	14.6	
Total	154	100

Table 12.2 Total social network users (2015)

Social network users by age (2015)

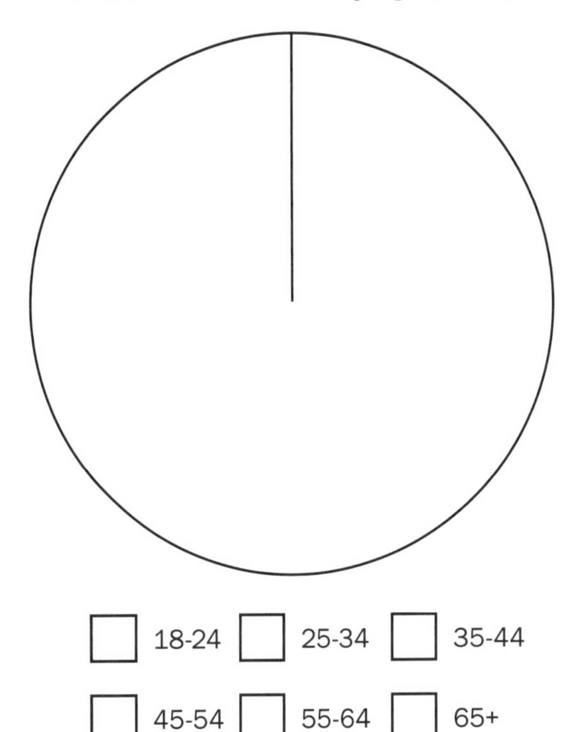

☐ 18-24 ☐ 25-34 ☐ 35-44

☐ 45-54 ☐ 55-64 ☐ 65+

4 Study Resource 12.2 and then complete the following activities.

a State the percentage of the world's population living in:

i Asia ______________ **iii** North America ______________

ii Africa ______________ **iv** Australia ______________

b State the percentage of the world's land area in:

i Asia ______________ **iii** North America ______________

ii Africa ______________ **iv** Australia ______________

c What conclusion can be drawn about the relative population densities of Asia and Australia?

____

____

____

____

ISBN: 9780170368155

5 Use the data in Table 12.3 to construct a percentage bar graph showing global coffee production.

Producer	%
Brazil	32
Vietnam	15
Indonesia	9
Colombia	8
Ethiopia	4
India	4
Rest of the world	28

Table 12.3 Global coffee production

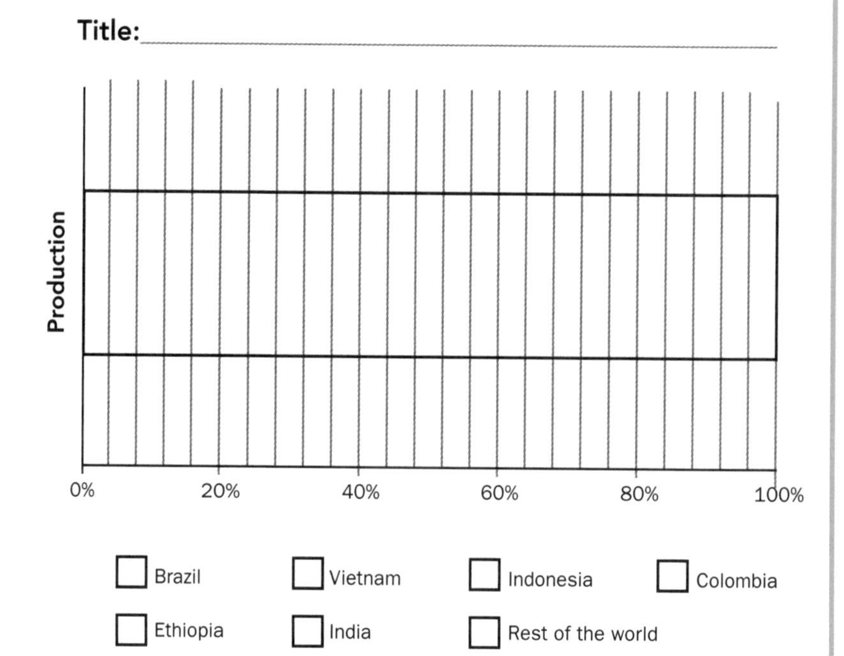

6 Use the data in Table 12.4 to construct a percentage bar graph showing the most common languages spoken around the world.

Language spoken	%
Chinese	21
English	6
Spanish	6
Hindu/Urdu	5
Arabic	4
Bengali	4
Portuguese	3
Russian	3
Japanese	2
German	2
Other languages	44

Table 12.4 Languages spoken around the world

Title:

Language

0% 20% 40% 60% 80% 100%

☐ Chinese ☐ English ☐ Spanish ☐ Hindu/Urdu ☐ Arabic ☐ Bangali

☐ Portuguese ☐ Russian ☐ Japanese ☐ German ☐ Other languages

ISBN: 9780170368155

13 Scatter graphs

In geography a scatter graph is used to represent the measurements of two related variables. It is particularly useful when one variable is thought to be dependent on the values of the other variable.

In a scatter graph, data points are plotted but not joined. The resulting pattern indicates the type and strength of the relationship between the two variables (Resource 13.1, 13.2).

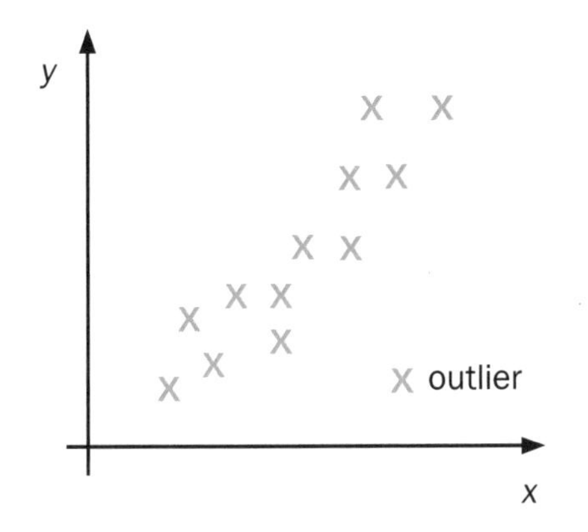

Resource 13.1 An example of a positive correlation

Resource 13.2 An example of a negative correlation

The line of 'best fit' or trend line is a straight line that is drawn on a scatter graph to show the general trend. Lines of best fit help identify the relationship between the two variables shown in the graph. A line of best fit that rises from left to right signifies a positive correlation (i.e. as variable *x* increases, so does variable *y*). A line of best fit that slopes down from left to right signifies a negative correlation (i.e. as variable *x* increases, variable *y* decreases). Data points that are not close to the line of best fit are called outliers.

To construct a scatter graph, follow the steps below:

1. Decide what information is to be plotted on each axis. In most cases, you will plot quantifiable data on both the *x*-axis and the *y*-axis.
2. Determine an appropriate scale for the two axes. Like line and bar graphs, scatter graphs also follow the graphing convention that requires the axes to follow a constant number scale starting from zero (e.g. 0, 5, 10, 15 or 0, 10, 20, 30). If the range of data is too broad, however, one axis may employ a logarithmic scale (Resource 13.3).
3. Having determined the range and scale of the data to be plotted, use a ruler to construct the axes. Like most graphs, scatter graphs have two axes: the *x*-axis is usually horizontal (i.e. runs across the bottom of the graph), while the *y*-axis is usually vertical (i.e. runs up the left side of the graph). Ensure both axes are long enough to accommodate the range of data you wish to show.
4. Label each axes (including the units of measurement) and give the graph a title that clearly states what it illustrates. If appropriate, the title should also include location and date specific information.
5. Next, plot the intersection of each value on the graph with an 'x'. It is important that you do not join the points together.
6. Draw a line of best fit.

ISBN: 9780170368155

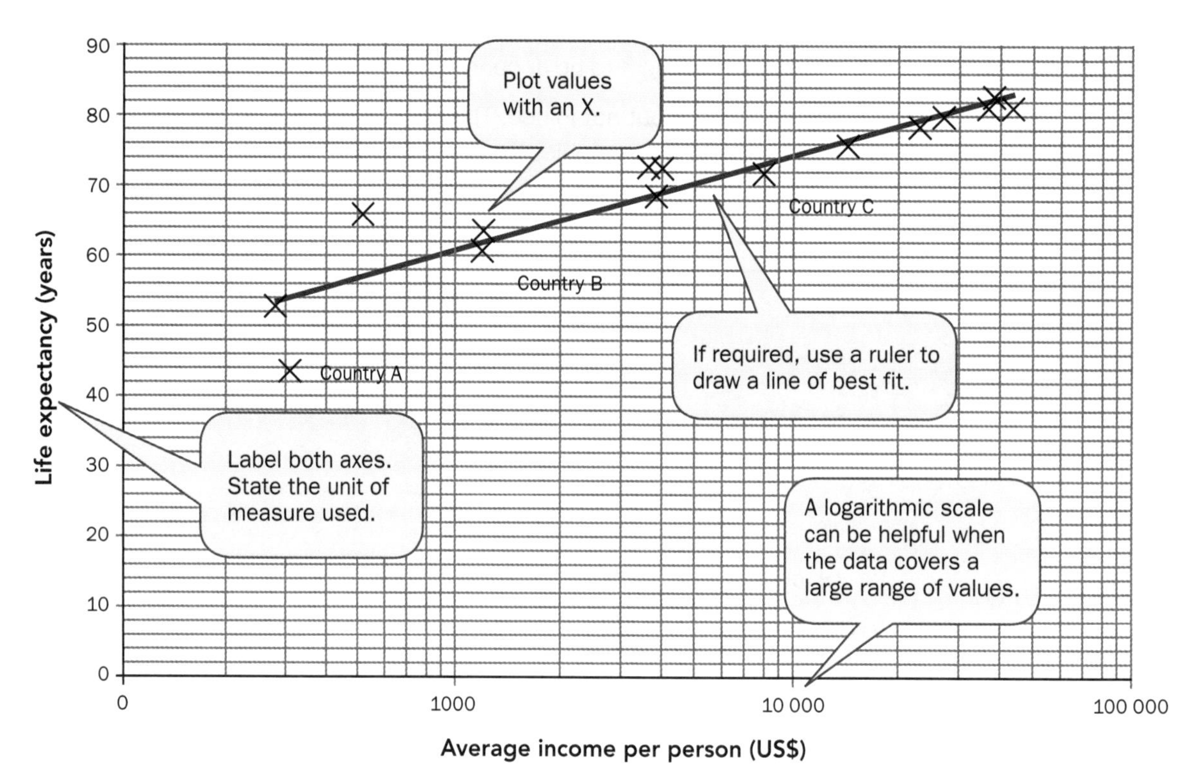

Resource 13.3

Learning Activities

1 Study Resource 13.3 and then complete the following activities.

a State the life expectancy and average income per person for:

i Country A ______

ii Country B ______

iii Country C ______

b Write a statement summarising the relationship between life expectancy and average income.

______

______

______

2 Using the data in Table 13.1:

a Construct a scatter graph to compare life expectancy with the average number of children per woman.

b Add a line of best fit to your graph.

ISBN: 9780170368155

Country	Children per woman (total fertility)	Life expectancy (years)
Timor-Leste	5.7	62
Ethiopia	4.1	62
Iraq	4.9	69
Mexico	2.2	74
Niger	7.6	58
New Zealand	2.0	81
Poland	1.2	77
Russia	1.7	71
Samoa	4.7	74
United Kingdom	1.9	83

Table 13.1 Life expectancy versus the number of children per woman (2013)

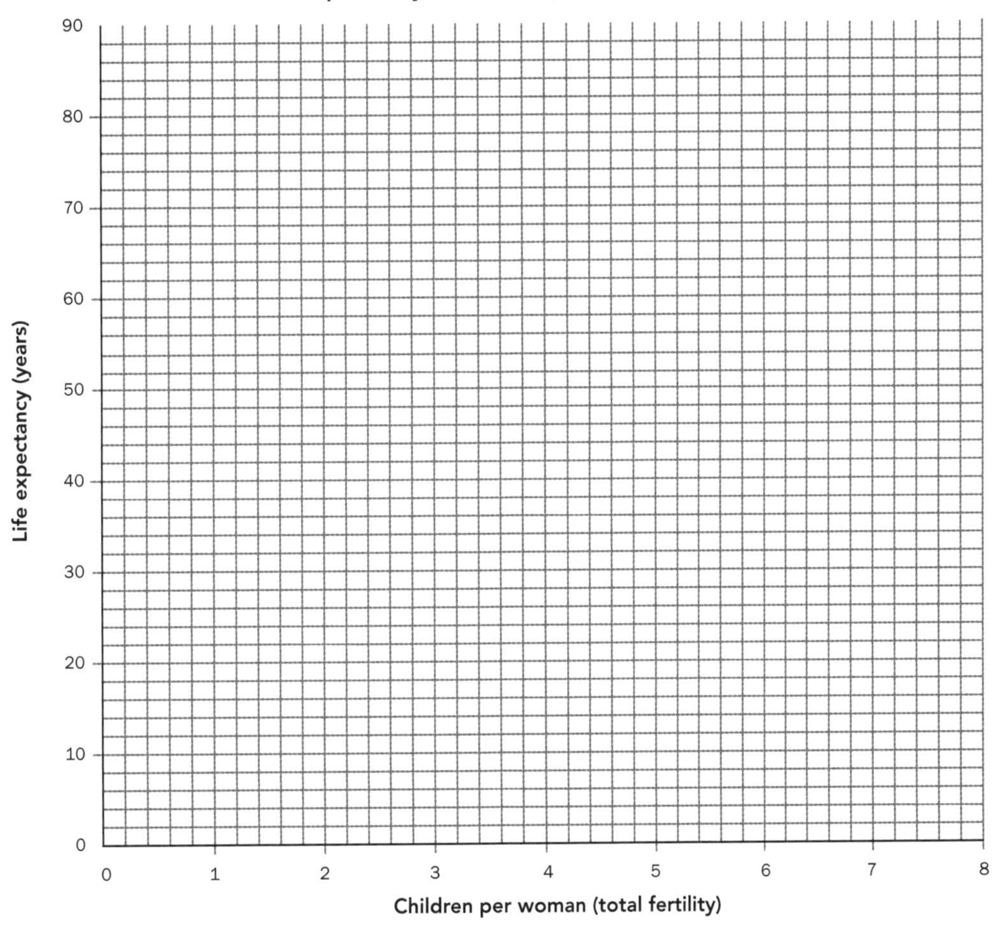

c Write a statement summarising the relationship between life expectancy and the average number of children per woman.

____

____

____

ISBN: 9780170368155

Learning Activities

3 Using the data in Table 13.2:

a Construct a percentage scatter graph to compare the number of Internet users per 100 and the percentage of urban population for selected countries.

b Add a line of best fit to your graph.

Country	Urban population (% of total)	Internet users (per 100 people)
Netherlands	81	84
Papua New Guinea	13	2
China	42	16
Indonesia	50	6
Italy	68	54
United Kingdom	90	72
Pakistan	36	11
Brazil	85	35
Spain	77	52
New Zealand	86	70

Table 13.2 Internet users (per 100) and the percentage of urban population

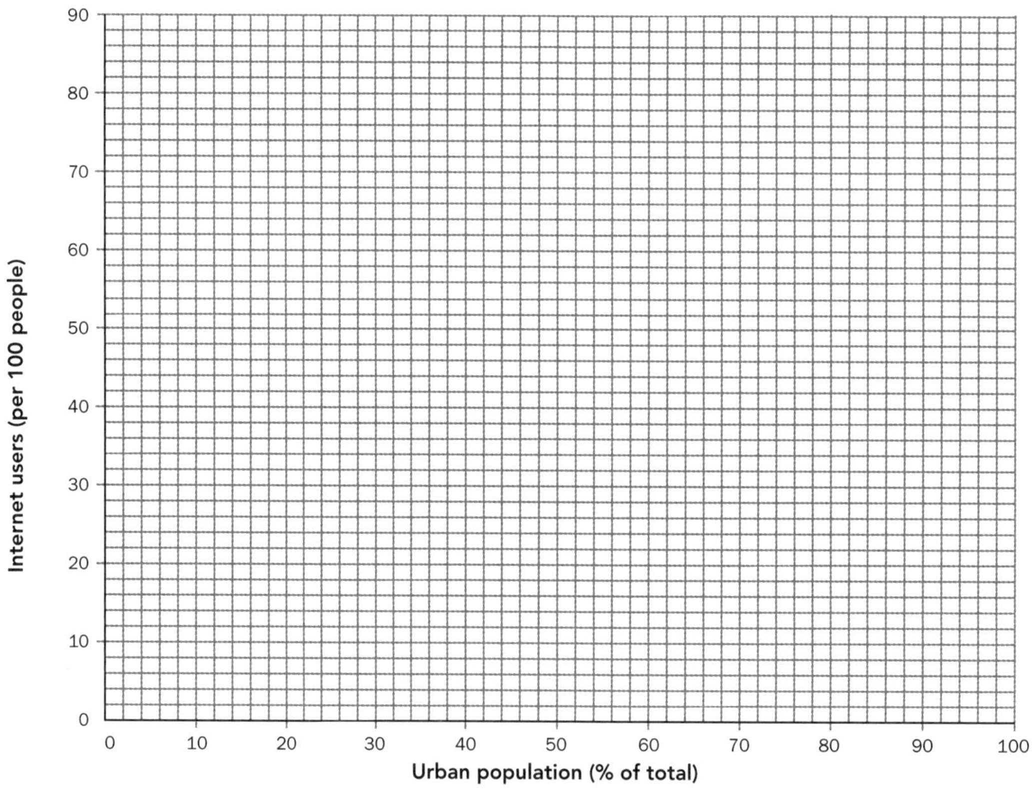

c Write a statement summarising the relationship between the number of Internet users per 100 and the percentage of urban population for selected countries.

____

____

ISBN: 9780170368155

14 Pictographs

A pictograph is a pictorial representation of statistical information, where each value is represented by a proportional number of pictures or symbols. A pictograph has a form similar to that of a bar graph.

To construct a pictograph, follow the steps below:

1. Decide what information is to be plotted on each axis. In most cases you will plot the non-quantifiable variable on one axis (e.g. country names, age groups, or periods such as months or years), and quantifiable data on the other.
2. Determine an appropriate scale for the variable axis. Like bar graphs, pictographs abide by the graphing convention that requires the variable axis to follow a constant number scale starting from zero (e.g. 0, 5, 10, 15 or 0, 10, 20, 30).
3. Having determined the range and scale of the data to be plotted, use a ruler to construct the axes. Like most graphs, pictographs have two axes: the *x*-axis is usually horizontal (i.e. runs across the bottom of the graph), while the *y*-axis is usually vertical (i.e. runs up the left side of the graph). Ensure both axes are long enough to accommodate the range of data you wish to show.
4. Label each axis (including the units of measurement) and give the graph a title that clearly states what it illustrates. If appropriate, the title should also include location and date specific information.
5. Choose a symbol to represent the variable you want to plot. Then decide the quantity that the symbol will represent. In Resource 14.1 🌍 represents one Earth. When plotting the symbols onto your graph, ensure they are drawn with constant spacing and width.
6. If appropriate, label each bar and include a key.

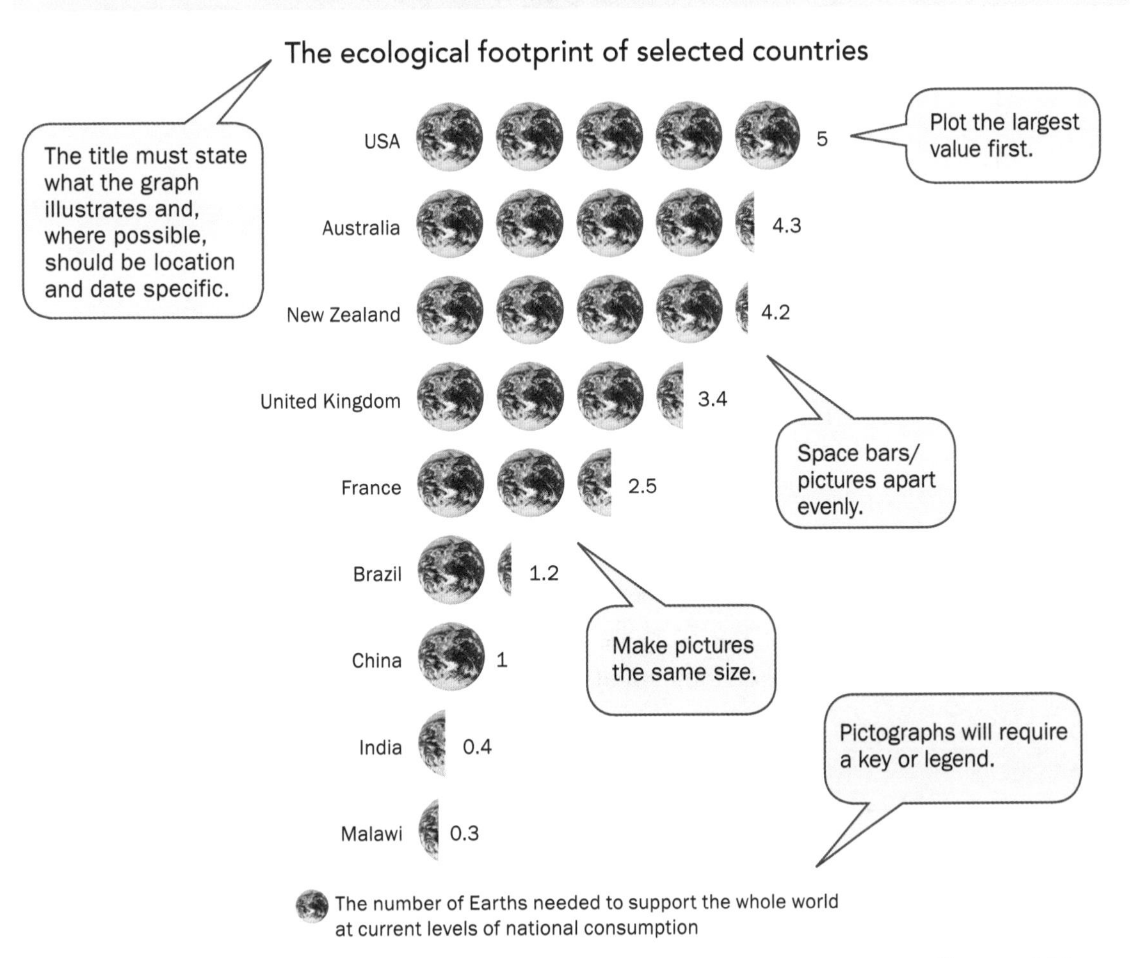

Resource 14.1

ISBN: 9780170368155

Learning Activities

1 Study Resource 14.1 and then complete the following activity.

a State the ecological footprint of:

i Australia ______________________

ii France ______________________

iii China ______________________

iv Malawi ______________________

2 Using the data in Table 14.1, construct a pictograph to show the world's top consumers of coffee.

	Coffee consumption (kg per person per year)
Finland	12
Italy	5.9
France	5.8
Brazil	4.8
New Zealand	3.7
Ethiopia	3.3

Table 14.1 Annual coffee consumption per person

3 Using the data in Table 14.2:

a Construct a pictograph to show the ten largest oil producing countries.

b Locate the countries listed in Table 14.2 on a world map and write a statement summarising the global distribution of the world's largest oil producing nations.

____

____

____

____

ISBN: 9780170368155

	Million bbl/day
Russia	9.9
Saudi Arabia	8.1
USA	5.0
Iran	4.1
China	4.0
Canada	3.2
Mexico	2.6
Iraq	2.4
Norway	2.3
United Arab Emirates	2.3

Table 14.2 Top oil producing nations

Top ten oil producing nations

Russia
Saudi Arabia
USA
Iran
China
Canada
Mexico
Iraq
Norway
United Arab Emirates

Million barrels per day (bbl/day)

4 Using the data in Table 14.3:

a Construct a pictograph to show the dairy cattle numbers by region.

b Write a statement summarising the regional distribution of dairy cattle.

Region	Dairy cows (000s)
Northland	284
Auckland	111
Waikato	1,164
Bay of Plenty	197
Gisborne	10
Hawke's Bay	113
Taranaki	493
Manawatu-Wanganui	220
Wairarapa	169
West Coast	150
Canterbury	878
Otago	251
Southland	550
Nelson-Marlborough	88

Table 14.3 Dairy cattle numbers by region (2014)

Dairy cattle numbers by region (2014)

Northland
Auckland
Waikato
Bay of Plenty
Gisborne
Hawke's Bay
Taranaki
Manawatu-Wanganui
Wairarapa
West Coast
Canterbury
Otago
Southland
Nelson-Marlborough

100,000 dairy cows

ISBN: 9780170368155

15 Climate graphs

A climate graph shows average temperature and rainfall experienced at a particular location throughout the year. It consists of a bar graph showing average monthly rainfall and a simple line graph showing average monthly temperature.

Climate graphs are constructed using climate data collected over several decades by organisations such as the New Zealand MetService and the National Institute of Water and Atmospheric Research (NIWA). These organisations in turn make the data available to others who rely on accurate climate information to make informed decisions. Viticulturists (wine makers) use climate data to assess frost risk for their crops, for example.

Interpreting climate graphs help geographers to draw conclusions about climate at a particular location. For instance:

- If the climate graph shows temperature increases in the middle of the year, then it is likely that the graph represents a location in the northern hemisphere. If the temperature decreases in the middle of the year then it is likely the climate graph represents a location in the southern hemisphere. If there is little variation in temperature during the course of the entire year, then it is likely that the climate graph represents a location within the tropics, near the equator.
- Locations within the tropics experience six months of cool-dry weather and six months of hot-wet or monsoonal weather. Therefore, if the climate graph shows a period of hot temperatures (25˚C or more), coinciding with a period of high rainfall, followed by a period of low rainfall and cooler temperatures (25˚C or less), then it is likely the climate graph represents a location in the tropics. However, if more than 60% of the rainfall occurs during the cooler months (i.e. the winter season), then it is likely the climate graph represents a location outside the tropics.
- Places that experience less than 25 mm of rain per month for the majority of the year are likely to be deserts irrespective of the location's temperature range.

To construct a climate graph, follow the steps below:

1. A climate graph has one horizontal axis banded by two vertical axes. To construct the horizontal axis, divide the axis into twelve even segments to represent the months of the year.
2. Place the rainfall scale (mm) on the left side of the graph and the temperature scale (˚C) on the right side of the graph. Determine an appropriate scale for the two vertical axes and follow a constant number scale starting from zero for both axes (e.g. 0, 5, 10, 15 or 0, 100, 200, 300).
3. Label all axes (including the units of measurement) and give the graph a title that clearly states what it illustrates. The title should include location specific information as well as the location's latitude and longitude.
4. Plot rainfall data for each month and colour each plotted bar blue.
5. Plot the average temperature data for each month with an 'x'. Ensure that each temperature value is positioned at the centre of the grid for each month. Join the data points with a red line in a smooth curve.

ISBN: 9780170368155

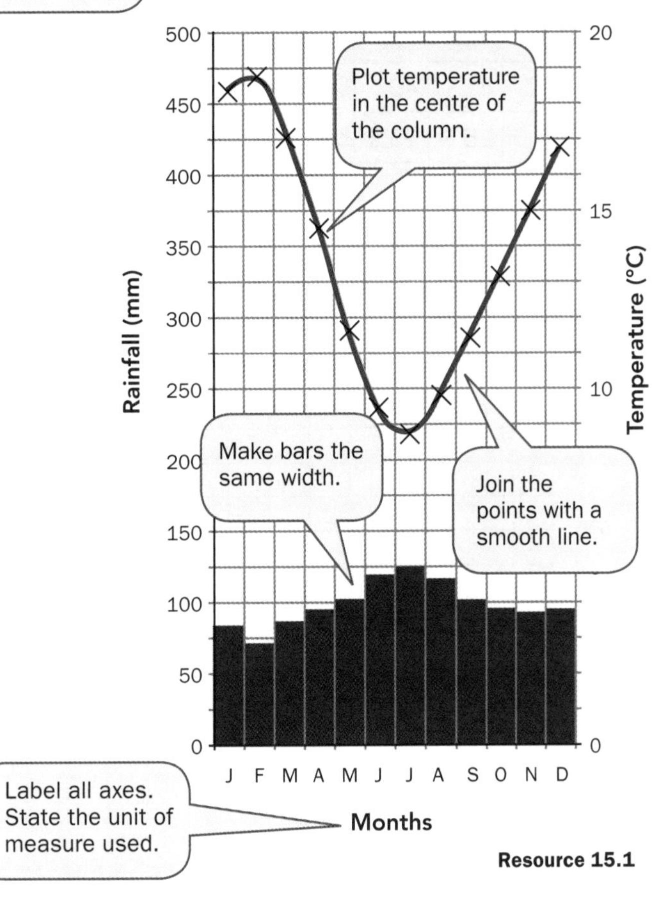

Resource 15.1

Learning Activities

1 Study Resource 15.1 and then complete the following activity:

a State the average temperature and rainfall for:

i January ______

ii April ______

iii July ______

iv October ______

b Identify the wettest and driest months of the year. ______

c Identify the warmest and coolest months of the year. ______

d Write a statement summarising Hamilton's annual climate.

______

______

______

______

ISBN: 9780170368155

Learning Activities

2 Using the data in Table 15.1, construct a climate graph for Suva, Fiji.

	J	F	M	A	M	J	J	A	S	O	N	D
Rainfall (mm)	319	271	413	401	255	141	112	140	163	211	254	264
Temperature (˚C)	26.1	26.3	25.9	25.1	24	23.2	22.3	22.4	22.7	24	24.6	25.1

Table 15.1 Climate data for Suva, Fiji

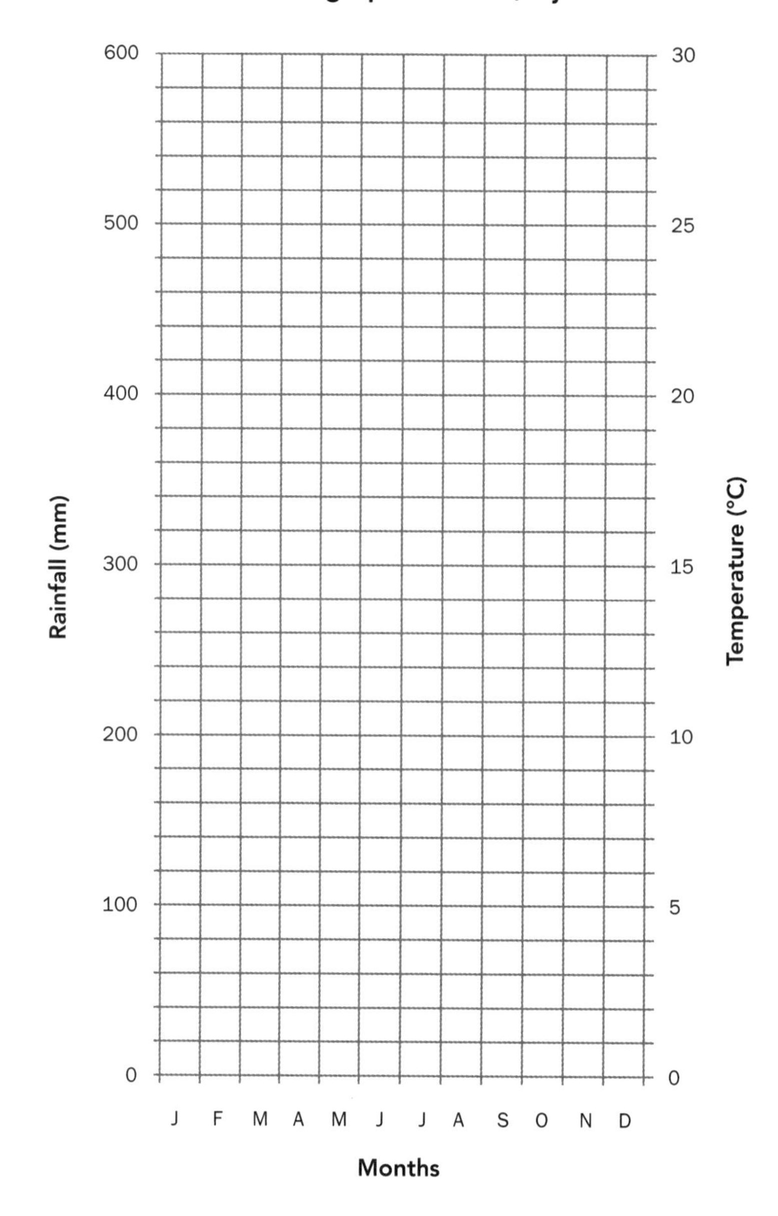

3 What evidence does the climate graph for Suva contain to suggest that Fiji experiences a cool-dry season and a hot-wet season?

ISBN: 9780170368155

16 Population pyramids

A population pyramid is a special type of horizontal bar graph that shows the age-sex structure of a population in one-year, five-year, or ten-year age groups.

The shape of a population pyramid reflects the influence of births, deaths and migration on a population over time, and shows whether a population is expanding, stable or likely to decline (Resource 16.1). In general a population with a high birth rate and low death rate will have a broad based, triangular-shaped pyramid. Populations with low birth rates and low death rates are usually narrower at the base and have straighter sides.

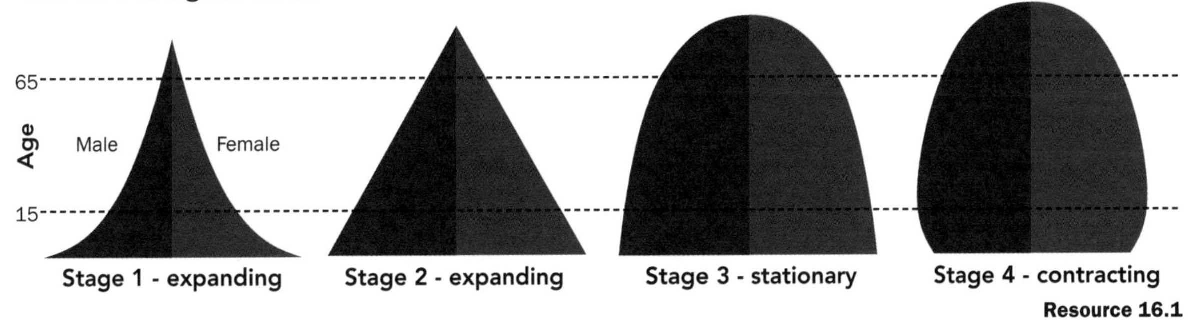

Resource 16.1

To construct a population pyramid, follow the steps below:

1. Use a ruler to construct the axes. Population pyramids have two axes: the horizontal or *x*-axis usually runs across the bottom of the graph, while the vertical or *y*-axis usually runs up the left side or centre of the graph.
2. Divide the vertical axis into segments that correspond with the age data you are using. Most population pyramids use one, five or ten-year age groups.
3. Determine an appropriate scale for the horizontal axis. It is recommended that you construct population pyramids using percentages rather than numbers, since this makes it possible to compare countries with different sized populations.
4. Label each axes and give the graph a title that clearly states what it illustrates. The title should also include location and date specific information.
5. Beginning at the bottom of the graph, plot the percentage of the population that is 0–4 years and male. Shade this bar on the pyramid and repeat using data for females, using a different colour. Repeat this step for each age group until the pyramid is complete (Resource 16.2).

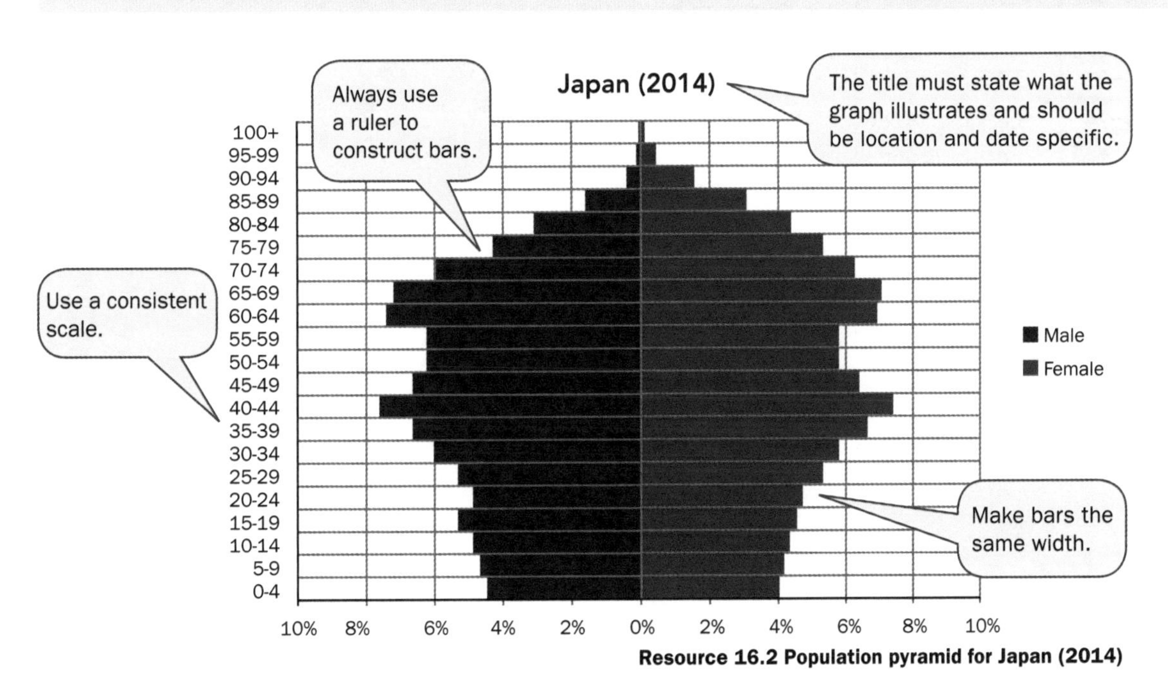

Resource 16.2 Population pyramid for Japan (2014)

ISBN: 9780170368155

Learning Activities

1 Study Resource 16.2 and then complete the following activities.

a Estimate the percentage of Japanese under the age of 15 years in 2014.

b Estimate the percentage of Japanese over the age of 65 years in 2014.

c Using Resource 16.1 as a guide, describe the shape of Japan's population pyramid.

2 Use the data in Table 16.1 to construct a population pyramid for the United Kingdom.

	Male	Female
0–4	6.2%	5.9%
5–9	6.1%	5.7%
10–14	5.6%	5.2%
15–19	6.1%	5.7%
20–24	6.9%	6.6%
25–29	7.1%	6.8%
30–34	6.8%	6.4%
35–39	6.2%	5.8%
40–44	7.1%	6.6%
45–49	7.6%	7.4%
50–54	7.1%	7.1%
55–59	6.1%	6.0%
60–64	5.3%	5.5%
65–69	5.4%	5.7%
70–74	3.8%	4.2%
75–79	2.9%	3.5%
80–84	2.0%	2.7%
85–89	1.1%	1.8%
90–94	0.5%	1.0%
95–99	0.1%	0.3%
100+	0.0%	0.1%

Table 16.1 Population, by age and sex for United Kingdom (2014)

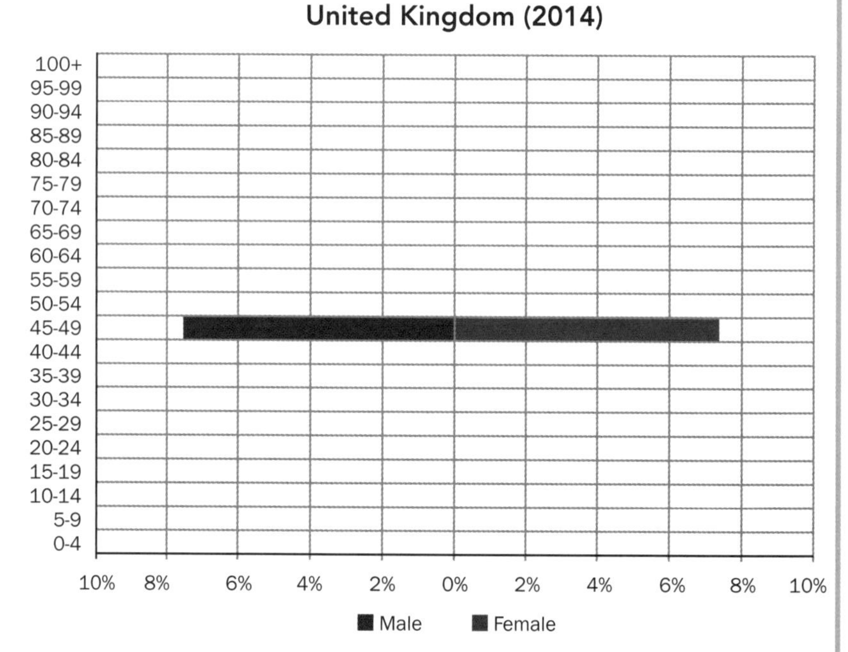

ISBN: 9780170368155

Learning Activities

3 What evidence does the population pyramid for the United Kingdom contain to suggest that it has an ageing population?

4 Use the data in Table 16.2 to construct a population pyramid for Niger.

	Male	Female
0–4	19.8%	19.5%
5–9	16.8%	16.6%
10–14	13.5%	13.4%
15–19	10.3%	10.5%
20–24	7.8%	8.2%
25–29	6.3%	6.7%
30–34	5.4%	5.6%
35–39	4.6%	4.6%
40–44	3.8%	3.7%
45–49	3.1%	3.0%
50–54	2.5%	2.4%
55–59	1.9%	1.8%
60–64	1.4%	1.3%
65–69	1.0%	1.0%
70–74	0.8%	0.8%
75–79	0.5%	0.5%
80–84	0.3%	0.3%
85–89	0.1%	0.1%
90–94	0.0%	0.0%
95–99	0.0%	0.0%
100+	0.0%	0.0%

Table 16.2 Population by age and sex for Niger (2014)

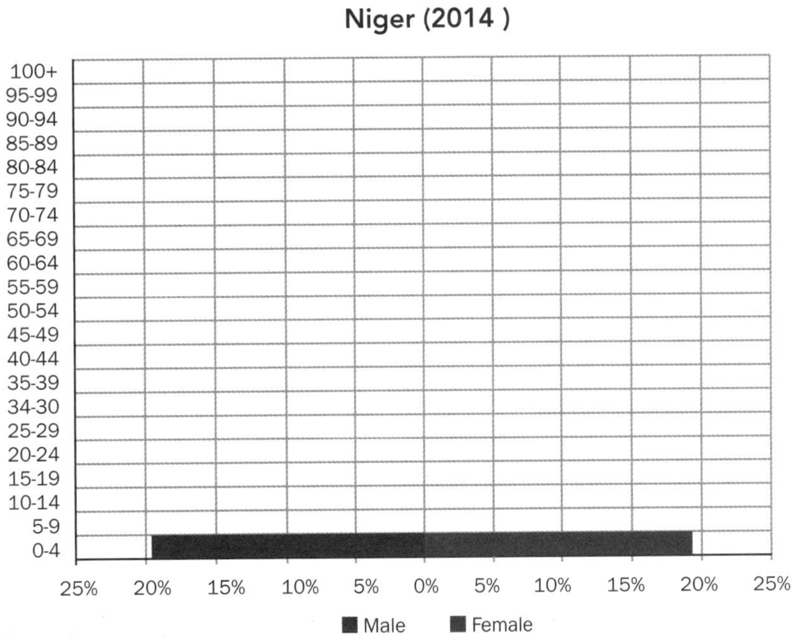

Niger (2014)

5 What evidence does the population pyramid for Niger contain to suggest that it has a youthful population?

ISBN: 9780170368155

Visual image interpretation

Visual images may be used to illustrate almost any aspect of geography. As a student of geography, you need to be able to identify, describe and interpret the geographical processes and ideas that can be seen in a visual image. You may be asked to identify and interpret what you can see, or you may be asked to compare the information from an image with that from a map.

17 Photograph interpretation

Geographers who record information while undertaking geographic research regularly use photographic images. Photographic images are an effective tool as they provide a visual record of natural and cultural features of the environment and help us to understand the way different elements of the environment relate.

Photographs also allow geographers to study how environments change over time. By comparing photographs taken at different times it is possible to analyse change in any one environment.

Three types of photographs are useful to geographers:

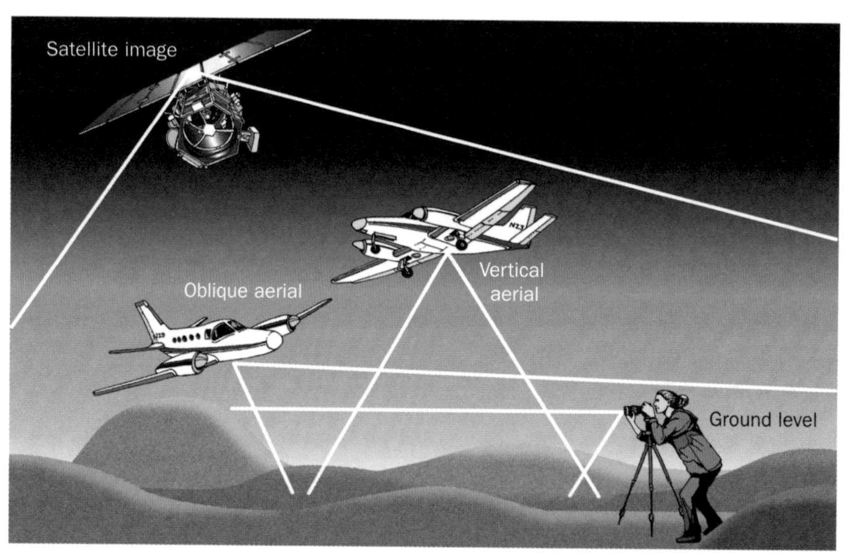

Resource 17.1 Types of photographs

- **Ground level photographs** are taken from the ground to maximise the horizontal view. With ground level photographs, foreground features appear larger than background features.
- **Aerial photographs** are taken of the Earth's surface from the air. They show a bird's eye view from either directly above (vertical) or from an oblique angle. Oblique photographs have an advantage in that they show both the tops and sides of objects, making them easier to identify. The main disadvantage of oblique photographs is that they do not have a consistent scale.
- **Satellite images** are created using data collected from satellites orbiting the Earth. With satellite photographs, spatial patterns are clearly visible over a large area (Resource 17.1).

ISBN: 9780170368155

Resource 17.2 Satellite image of Venice, Italy

To interpret a photograph, follow the steps below:

1. Determine whether the photograph shows an aerial or ground level perspective. If the perspective is aerial, determine the orientation of the photograph (i.e. is it vertical or oblique?).
2. Identify any visual clues that help identify the photograph's setting and location.
3. Identify the main natural features (e.g. relief and drainage features, vegetation, soil and climate) and cultural features (e.g. patterns of settlement, transportation networks and other land uses) of the environment shown in the photograph.
4. Examine the way different features shown in the photograph interact. The photograph might show settlement along a coastline, for example, or on the flood plain of an adjacent river.

Learning Activities

1 Study the photographs in Resources 17.3, 17.4 and 17.5. Identify which photograph is:

a An oblique aerial photograph

____

b A vertical aerial photograph

____

c A ground level photograph

____

Resource 17.3 Manhattan, New York

Resource 17.4 Manhattan, New York

Resource 17.5 Scarsdale, New York

ISBN: 9780170368155

Learning Activities

2 How do vertical aerial photographs differ from oblique aerial photographs?

3 Study Resource 17.6 and then complete the following activity:

Resource 17.6 Rio de Janeiro, Brazil

a Write a statement describing each of the following landscape features in relation to the landscape shown:

i Relief

ii Vegetation

iii Drainage (i.e. rivers and lakes)

iv Settlement

v Other land uses

b How do each of the features in Question 3a relate or interact with other features of the landscape?

4 Use the Internet to find ground level, vertical and oblique photographs of a city of interest to you.

ISBN: 9780170368155

18 Précis sketches

Geographers regularly use sketches to identify and record features of the landscape. If a sketch is drawn from a photograph it is called a précis sketch (which simply means a summary sketch).

You do not need artistic ability to draw an effective précis sketch. You do, however, need to be able to illustrate your understanding of the landscape shown in the photograph, by being able to identify and draw the important geographic features found within it.

To construct a précis sketch from a photograph, follow the steps below:

1. Study the photograph and choose the area to be included in the sketch.
2. Draw a frame in the same shape as the photograph you wish to sketch. If you are drawing a précis sketch from an oblique aerial photograph, then it may be beneficial to construct a trapezoid-shaped frame with a narrow base, to compensate for the narrow field of vision in the foreground.

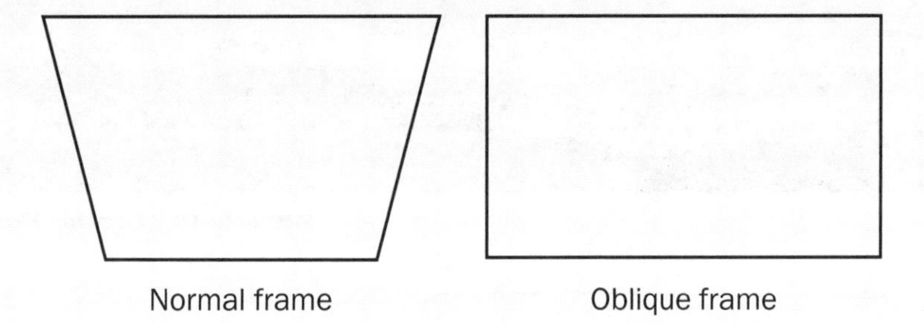

3. Divide both the photograph and the précis sketch frame into three equal areas: foreground, middle distance and background (Resource 18.1).

Resource 18.1

4. Select the main natural and cultural features shown in each area of the photograph, and sketch the outline of their shape into the frame of the précis sketch.
5. Use colour shading to highlight main geographical features.
6. Label the main geographical features.

ISBN: 9780170368155

Resource 18.2 Lambton Harbour, Wellington

Resource 18.3 Sample précis sketch of Lambton Harbour, Wellington

Do not include any moving objects like cars or people.

ISBN: 9780170368155

Learning Activities

1 Explain why geographers construct précis sketches from photographs.

2 Construct a précis sketch for Resource 18.4. Label the main geographical features illustrated.

Resource 18.4

ISBN: 9780170368155

Learning Activities

3 Construct a précis sketch for Resource 18.5. Label the main geographical features illustrated.

Resource 18.5

ISBN: 9780170368155

19 Cartoon interpretation

Geographers have long acknowledged the value of cartoons as a way of fostering a greater understanding of contemporary geographic issues.

Interpreting cartoons requires an appreciation of the various techniques that a cartoonist uses to convey or communicate an idea.

When interpreting cartoons in a geographical context, three questions should be asked of the cartoon:

1. What information does the cartoon convey? This includes identifying the contemporary geographic issue portrayed in the image, and the perspective or viewpoint of the cartoonist.
2. What geographical concepts or ideas are addressed in the cartoon?
3. What are the geographical implications of the issue addressed by the cartoonist?

A finished cartoon will usually contain a combination of elements (Resource 19.1).

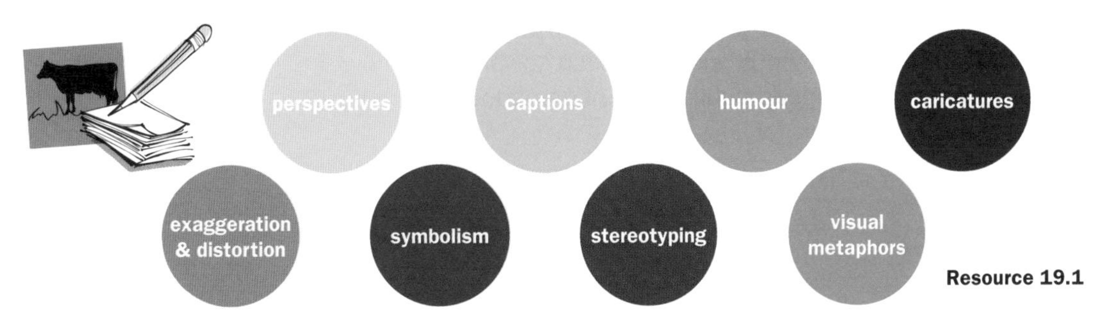

Resource 19.1

The following resource, illustrated by cartoonist Garrick Tremain, demonstrates how several techniques can be combined into one cartoon to communicate an idea of geographical importance.

Resource 19.2

Context: The cartoon refers to the environmental disaster in the Gulf of Mexico, which saw oil spill from a BP rig for more than two months, and relates the disaster to news that Gerry Brownlee, Minister of Energy and Resources, awarded a Brazilian oil company an exploration permit on the Raukumara block off New Zealand's East Coast. The minister had been quoted at the time as saying that he was 'aware of environmental concerns, but legislation was being worked on in terms of environmental requirements needed.'

ISBN: 9780170368155

Learning Activities

1 With the help of an online dictionary, write a definition for each of the following terms:

contemporary issue	
symbolism	
caricature	
stereotype	
visual metaphor	
perspective	

2 Study Resource 19.2 and complete the following activities.

a As a class, discuss the point of view being expressed in the Tremain cartoon. What does it tell us about our perception towards offshore oil exploration? What recent events have shaped public perceptions about oil exploration?

Your notes: ______________________________

b Working in groups, discuss the issue raised in the cartoon. Share the main points raised in your group's discussion with the rest of the class. To what extent do you think this cartoon reflects our views on environmental protection? Use this information to write a short statement explaining the point Tremain is trying to communicate in his cartoon.

Your notes: ______________________________

ISBN: 9780170368155

Learning Activities

3 Study the cartoon in Resource 19.3 and then complete the following activities.

Resource 19.3

Context: The cartoon shows a deep ravine dividing the land where the poor live from that of the rich. A newspaper reads 'NZ rich-poor gap widens fastest in world'. A fat rich man rejects these statistics and says the gap is not big and it is being bridged. A flimsy wire bridge with planks reading 'wages', 'asset sales', 'jobs' and 'cost of living' spans the gap.

a Identify the issue addressed in the cartoon.

b Describe the geographical processes relevant to the issue.

c Outline at least two different perspectives relevant to the issue addressed in the cartoon.

ISBN: 9780170368155

Learning Activities

d Identify the actions that individuals, groups and governments can take to address the issue highlighted by the cartoonist.

4 Study the cartoon in Resource 19.4 and then complete the following activities.

Resource 19.4

Context: The Alastair Nisbet cartoon published in 2011 shows several nineteenth century settlers observing 'volcanic cones', a 'swamp' and 'faultlines' and deciding to build a city. The cartoon refers to the way many New Zealand cities are built on or near volcanoes, faultlines and swamps.

a Who are the 'forefathers' referred to in the cartoon?

b Describe the perspective of the forefathers with regard to town planning.

c Name a New Zealand city that has been built in areas vulnerable to the threats posed by:

i volcanic activity

ii faultlines

iii swamp land

ISBN: 9780170368155

20 Annotating maps and diagrams

Examination and internal assessment questions frequently refer to labelling and annotating maps and diagrams. Annotating requires you to write relevant, appropriate comments on a map or diagram and to locate the comments in the relevant location.

To annotate a map or diagram, follow the steps below:

1. Identify the main features or important geographic ideas you wish to communicate to your audience.
2. Neatly write the annotation text on the map or diagram with a fine black pen.
3. Using a ruler, draw a line to link the annotation with the correct feature.

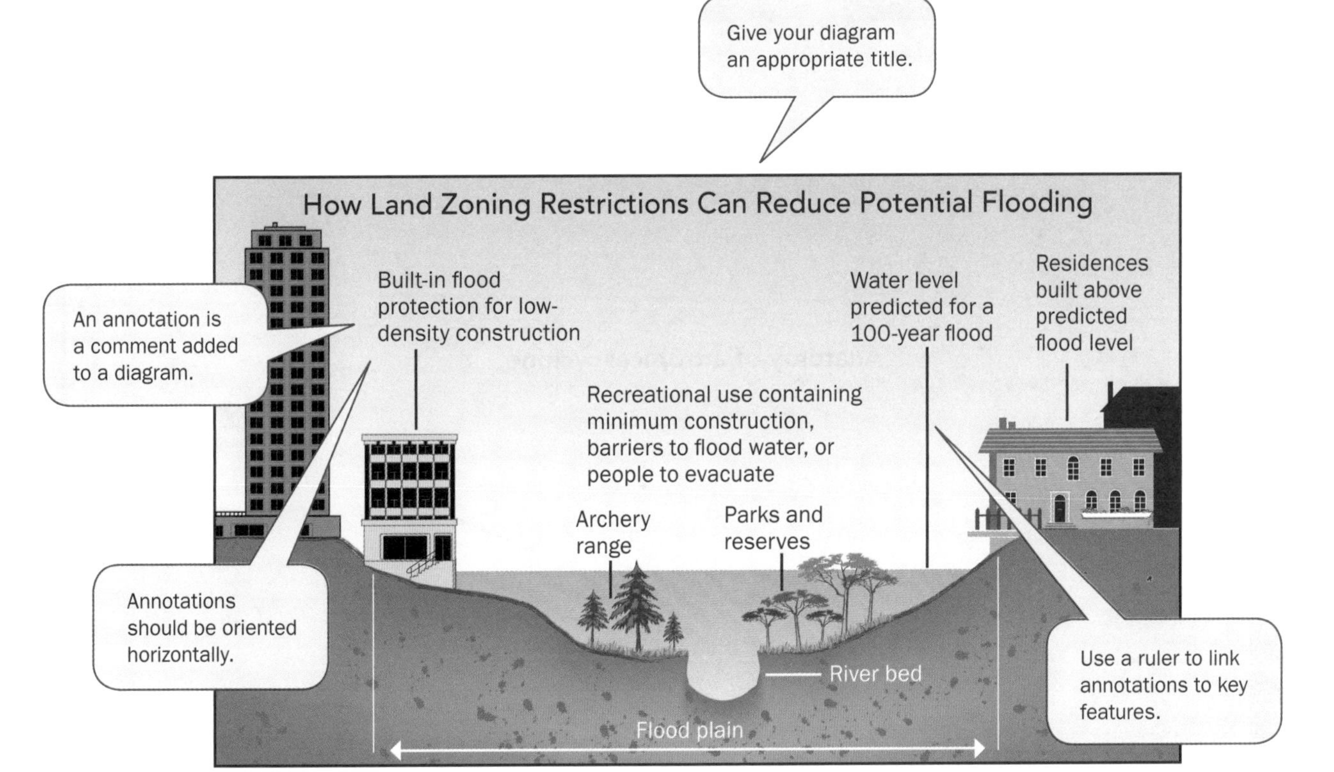

Resource 20.1 How land zoning restrictions can reduce potential flooding

Learning Activities

1. Explain how land zoning restrictions can reduce the potential effects of flooding.

ISBN: 9780170368155

Learning Activities

2 Use the information in the text boxes below to annotate the diagram showing the anatomy of a tropical cyclone (Resource 20.2).

Direction of movement	Strong updraught as warm moist air rises	Winds circulate in a clockwise direction (in the southern hemisphere)
Maximum wind speeds are located near the eye of the tropical cyclone	Ocean temperatures of at least 26°C provide a constant supply of warm moist air	Cooling and condensation leads to precipitation

Anatomy of a tropical cyclone

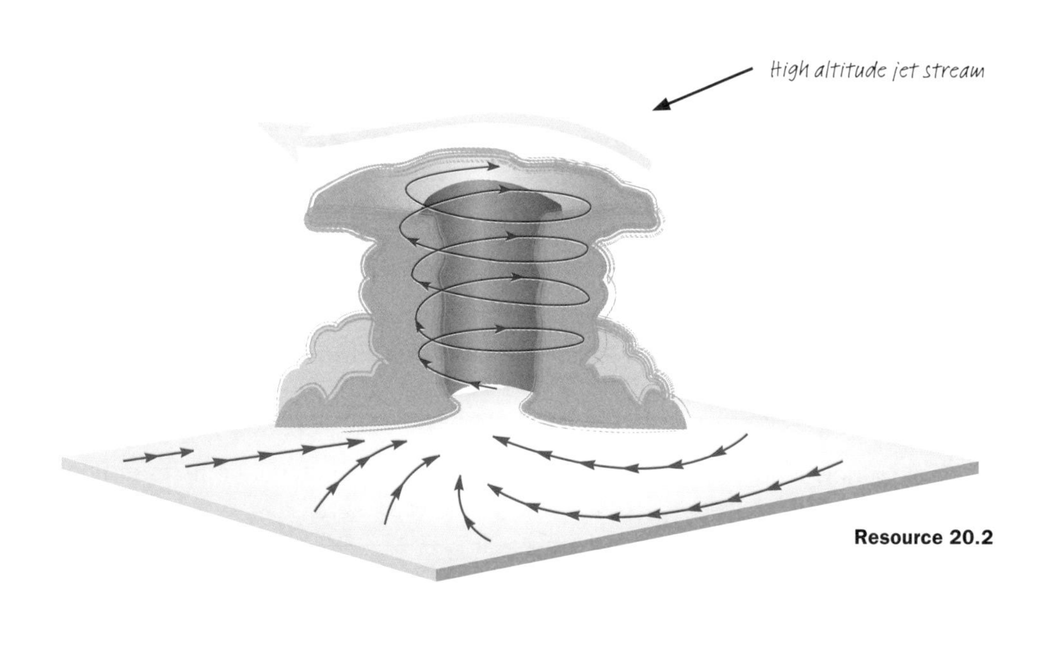

Resource 20.2

ISBN: 9780170368155

3 With the help of a map or atlas showing New Zealand's geological features, use the information in the text boxes to annotate the map of the North Island showing the location of its volcanic fields (Resource 20.3). An example of how to correctly annotate the diagram has been completed for you.

Taupo (dormant) : The lake is a giant caldera caused by a series of huge eruptions ending about 1850 years ago when everything around the vent for 20 000 sq km was destroyed.

Okataina (dormant): In the last 250 000 years there have been around five major eruptions. One of the most devastating was Tarawera in 1886.

White Island (active): A major eruption could potentially cause a tsunami threat to the Bay of Plenty coast.

Northland (dormant): The most recent eruption was Mt Te Puke (near Waitangi) about 1500 years ago.

Tongariro (active): This area includes Ngauruhoe, Ruapehu and Tongariro. Ruapehu erupted as recently as 2007, producing lahars. Ngauruhoe last erupted in 1975.

Taranaki (dormant): There have been at least nine eruptions in the last 500 years. The main hazard is ash.

The Location of North Island Volcanic Fields

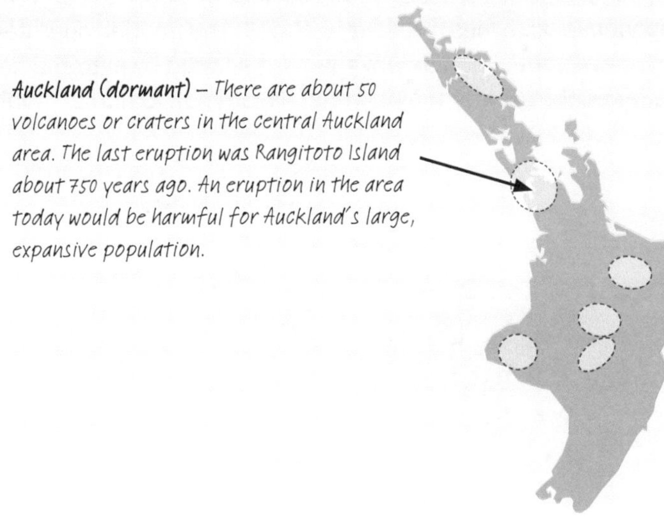

Auckland (dormant) *– There are about 50 volcanoes or craters in the central Auckland area. The last eruption was Rangitoto Island about 750 years ago. An eruption in the area today would be harmful for Auckland's large, expansive population.*

Resource 20.3

ISBN: 9780170368155

Geographic models

A model is a simplified representation of something that is real. You have most likely seen and used models in the past. A globe, for instance, is a model of Earth. Geographers use models to analyse geographic processes when the real object of study is too large to examine, or where the processes that created it operate over too long a period.

Geographical models range from simple conceptual models – such as box and arrow diagrams showing the flow of energy between components of a system – to complex computer-based mathematical models. Physical geographers construct physical models like stream tables, for example, to investigate the impact of river or fluvial processes on Earth's surface. Climate scientists, on the other hand, use complex mathematical models to predict changes in climate over time.

In this chapter you will learn about four conceptual models that are important to the study of Level 1 NCEA Geography.

21 Hazard risk model

The hazard risk model considers the probability (chance or likelihood) of an extreme natural event causing harm to social and economic activities (Resource 21.1). The model operates under the assumption that the larger an interaction between an environment shaped by extreme natural events and the vulnerability of its inhabitants, the greater the potential hazard risk to human health, property, economy and infrastructure.

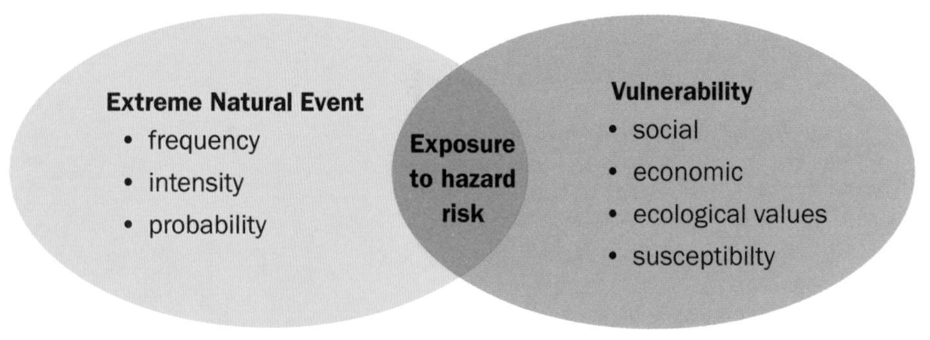

Resource 21.1 Hazard risk model

A number of factors can increase one's vulnerability to natural hazards. People are more vulnerable to the effects of natural hazards when there is:

- High population density. Large urban areas such as Wellington and Christchurch are especially vulnerable to natural hazards.
- Lack of understanding or awareness. People new to an area (e.g. immigrants) may be unaware of the hazard risk posed by the natural environment; people (individually or collectively) may not even be aware that they live or work in a hazard prone area, and consequently take no action to prepare for the likelihood of a hazard event.

ISBN: 9780170368155

- No early warning system or effective means of communication.
- Lack of preparedness (i.e. awareness education, emergency personnel, building codes).
- Inability to adequately respond to the hazard threat (i.e. lack of stable governance, emergency management, insurance cover).

Some of these factors help explain why poorer countries are more vulnerable to the effects of natural hazards than richer countries.

Learning Activities

1 The following table considers the relationship between the level of risk posed by a natural hazard, the likelihood of the event occurring, possible effects on human health and social and economic activities, and level of community preparedness.

Your task is to attempt to fill the table in with real case studies, examples and locations. For example, a high likelihood of a hazard event occurring with low predicted losses would be a volcanic eruption on White Island in the Bay of Plenty.

		Level of risk posed by a natural hazard		**Likelihood of the event occurring**		**Possible effects on human health and social & economic activities**	
		High	Low	High	Low	High	Low
Likelihood of the event occuring	High						
	Low						
Possible effects on human health and social & economic activities	High						
	Low			A volcanic eruption on White Island, Bay of Plenty.			
Level of community preparedness	High						
	Low						

Table 21.1

ISBN: 9780170368155

22 The wind rose

A wind rose is a diagrammatic tool used by geographers to show patterns in the frequency and direction wind blows from at a specific location. It shows the percentage or frequency of occurrence of winds at various strengths for the eight compass points. The length of the value point along each axis gives the overall frequency of wind from that particular direction.

In the example below for Kelburn, Wellington we can see that a little over 21% of wind samples show wind from the north. This is made up of 11% of speeds ranging from 1–19 km/h (light blue band), 6.5% in the 20–29 km/h (green) range, 2% of 30–39 km/h, and 2% exceeding 39 km/h (Resource 22.1).

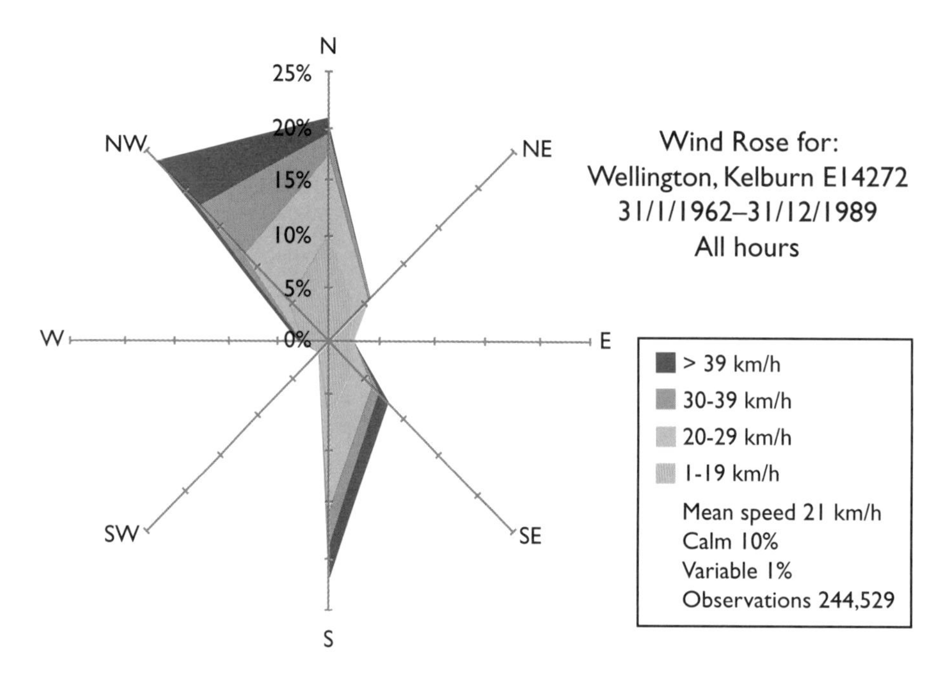

Resource 22.1 Historical wind rose for Kelburn, Wellington

The important thing to remember when describing wind patterns is that wind direction is always described according to the direction the wind blows *from* (Resource 22.2). A westerly wind blows *from* the west, for instance, and a southerly wind blows *from* the south.

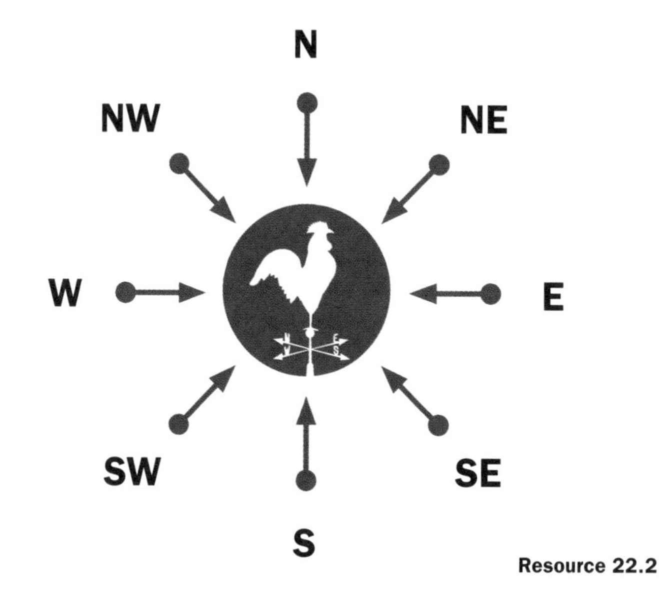

Resource 22.2

ISBN: 9780170368155

Learning Activities

1 Study the wind rose for Palmer Station, Antarctica and then complete the following activities.

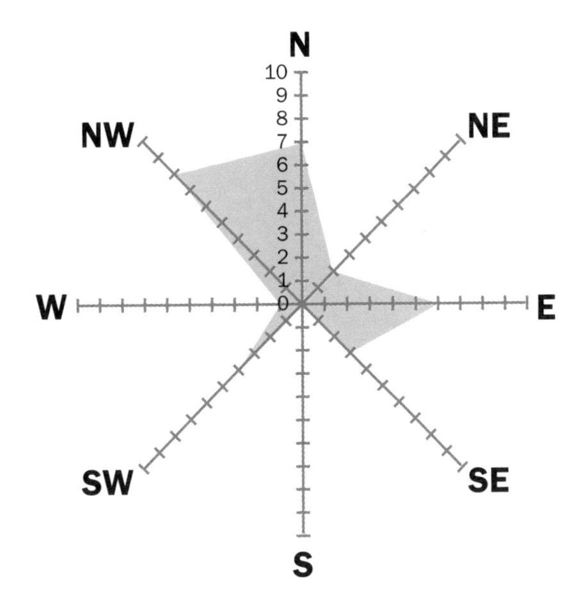

a State the number of days the wind at Palmer Station blew from the:

i North ______

ii North-east ______

iii South-west ______

iv West ______

b State the direction of the prevailing wind (i.e. single most frequent wind).

______

c Write a statement to describe the general pattern (frequency and direction) of wind at Palmer Station.

______

______

______

______

______

______

ISBN: 9780170368155

23 Demographic transition model

The demographic transition model is used in geography to describe the changes in natural population growth (i.e. birth and death rates) that accompany economic development (Resource 23.1). It is used by geographers to describe population change in industrialised countries and suggests that, given time, all countries pass through similar demographic changes.

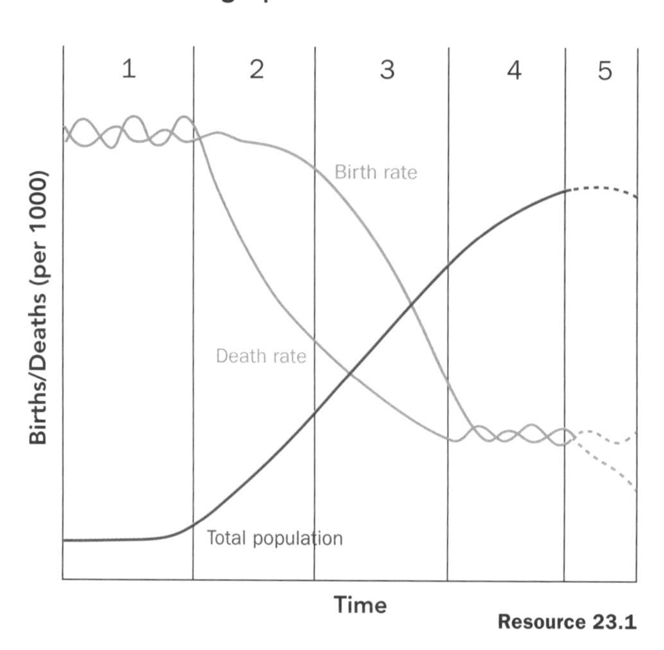

Resource 23.1

1 Using Resource 23.1 as a guide, draw a line to match each description with a stage of the demographic transition model.

Stage 1	Developed technology; more women in the workforce; low birth and death rates; natural increase very low; population increases very slowly or is static.
Stage 2	Higher living standards and education; birth control programmes; changing values reduces emphasis on large families; falling birth rate; population increases but not so fast.
Stage 3	Birth rate below replacement level; population decline.
Stage 4	New technology; improved health care and hygiene; falling death rate; static birth rate; rapid population growth.
Stage 5	Low technology; high birth rate; high death rate; small natural increase

ISBN: 9780170368155

2 Use the data in Table 23.1 to construct a multiple line graph to show population transition in Sri Lanka.

Year	Crude birth rate	Crude death rate
1911	38	31
1921	39	27
1931	38	22
1936	34	22
1940	36	21
1946	39	19
1953	39	11
1963	35	9
1971	30	8
1976	28	8
1983	26	6
1996	20	5
2006	16	7
2011	17	6

Table 23.1 Population transition in Sri Lanka

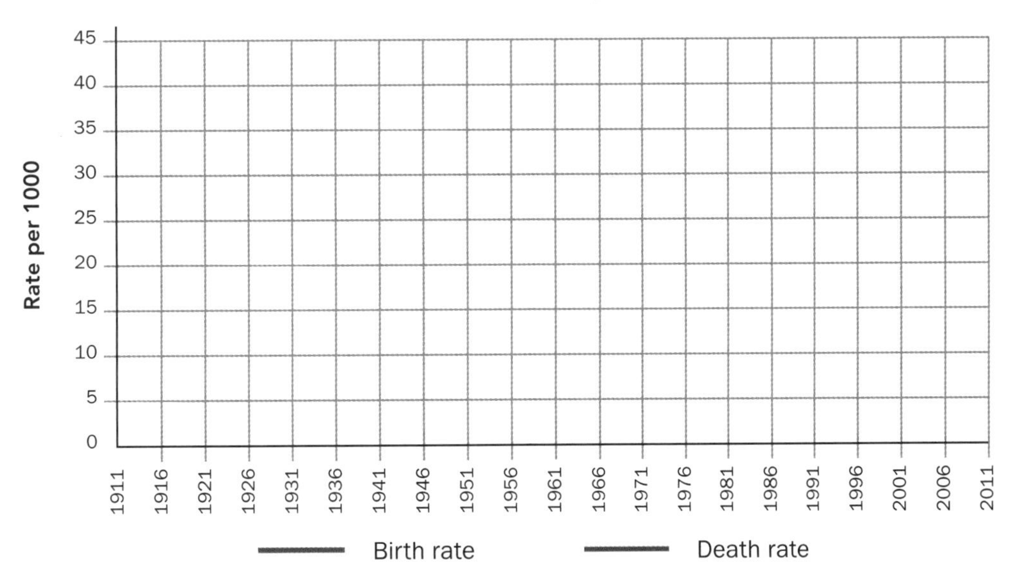

3 State the years that Sri Lanka was in:

a Stage 1 ______________________

b Stage 2 ______________________

c Stage 3 ______________________

4 Has Sri Lanka reached Stage 4 of the demographic transition model? Using evidence from your graph, justify your answer.

____

____

____

____

ISBN: 9780170368155

24 Lee's migration model

One of the most common models used to describe the movement of people from one country or region to another is Lee's migration model (Resource 24.1). Lee described migration in terms of push and pull factors in the places of origin and destination, and the intervening factors between them.

According to the model, potential migrants will decide whether to migrate based on their perception of the destination relative to their current place of residence. If a potential migrant's perception of the destination is more positive than that of their current residence, then migration is likely to occur. Any intervening obstacles (barriers to migration) have to be considered, however, before any move can take place. Examples of intervening obstacles include the financial cost of moving, language barriers and residency requirements at the destination.

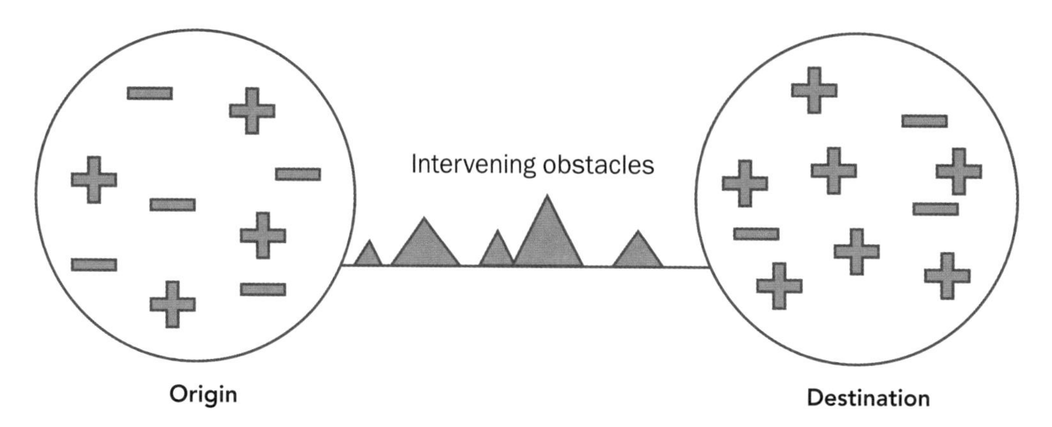

Resource 24.1 Lee's migration model

Learning Activities

1 Lee's model divides the factors causing migrations into two groups: push (-) and pull (+) factors. Push factors are negative things perceived about the country or region that one lives in, and pull factors are positive things that attract migrants to another area.

- Not enough jobs
- Few opportunities
- Stable governance
- Quality education
- Better medical care
- Attractive climates
- Security
- Primitive living conditions
- Desertification
- Famine or drought
- Political fear or persecution
- Slavery or forced labour
- Poor medical care
- Loss of wealth
- Natural disasters
- Job opportunities
- Better living conditions
- Political or religious freedom
- Civil war
- Lack of political or religious freedom
- Pollution
- Unaffordable housing
- Discrimination
- Unsafe housing
- Family links
- Better chances of marrying

ISBN: 9780170368155

Learning Activities

From the list of factors on the previous page, decide which are push or pull factors:

Push factors	Pull factors

ISBN: 9780170368155

Fieldwork and social skills

25 Interpreting and completing a continuum to show value positions

Perception is the term used to describe the way people make judgements or form opinions about the world around them. Many things can influence your perception but some of the most important influences include our personal values, spiritual beliefs, and customs and traditions. Because what influences us as individuals varies from person to person, or from group to group, people perceive the world around them differently.

The funnel below illustrates the way different influences shape our perception of the world.

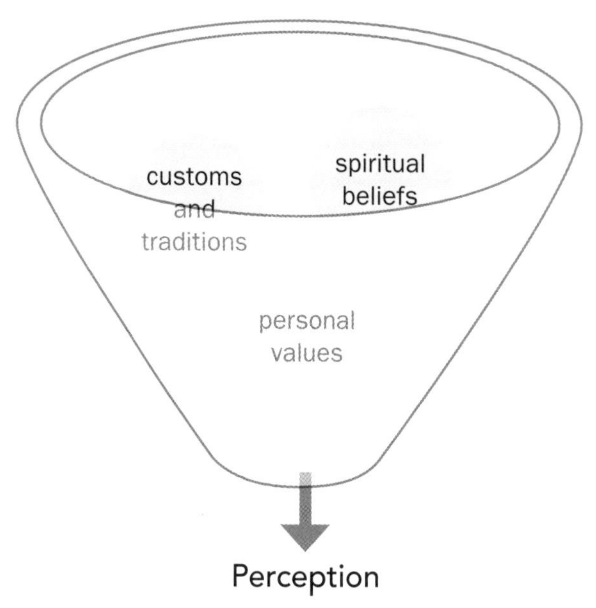

Resource 25.1

A values continuum is a simple yet effective tool used by geographers to display viewpoints on a particular issue. Completing a values continuum involves plotting individual value positions along a line representing varying degrees of commitment to a value. In the following example, a group of Year 11 students was asked how they viewed themselves in relation to the wider natural environment. They did this by placing an 'x' in a location along the continuum they believed best represented their personal viewpoint.

We are a part of the natural environment. We must aim to protect the Earth's finite resources at all costs. Consumerism exploits the environment.

The environment exists for our exclusive use. Resources should be used to maximise our standard of living. Environmental problems will be resolved by technological advancement.

ISBN: 9780170368155

Learning Activities

1 Using Resource 25.1 as a guide, describe the way your personal values, beliefs, customs or traditions have influenced your perception of the natural environment.

2 Put a cross on the values continuum on the previous page to show your viewpoint on the issue. How does your viewpoint of the natural environment differ to that of your:

a Closest friend?

b Geography teacher?

c Parents?

d Grandparents?

3 Now that you have identified what has shaped your perception of the natural environment, where do you stand on the following viewpoints about the environment?

Statement	**Strongly agree**	**Agree**	**Neither agree nor disagree**	**Disagree**	**Strongly disagree**
The natural environment only exists to provide resources for people to improve their quality of life.					
Recycling is only worth it if you have to walk no further than your front gate.					
It does not matter if you drop litter because other people are paid to clear up the mess.					
If we pollute the environment, science will invent new technological solutions to these problems.					
It is better to travel by car because cars are more reliable, more flexible and more convenient. Environmental effects are irrelevant.					

Continued on next page

ISBN: 9780170368155

Learning Activities

Statement	Strongly agree	Agree	Neither agree nor disagree	Disagree	Strongly disagree
If the environment is in trouble, the government should do something about it.					
My grandparents did not worry about the environment they were leaving for me, so why should I worry about the environment I leave my grandchildren?					
I have no idea what happens to the rubbish after the refuse collectors take it away.					
Where you live influences what you think about the environment.					

4 A two-dimensional continuum can be used to explore two related issues at the same time. Put a cross on the values continuum below to show your viewpoint on each issue. Indicate on the continuum how your viewpoint of the natural environment differs to that of your:

a Closest friend

b Geography teacher

c Parents

d Grandparents.

Supports preservation of the natural environment regardless of cost.

Caring for the environment is our collective responsibility.

Caring for the environment is a matter of individual choice.

Against preservation of the natural environment at any cost.

ISBN: 9780170368155

26 Conducting questionnaire surveys

A questionnaire survey is a research method used in human geography to gather information about the perceptions, spatial interactions and behaviours of a population, by asking standardised questions to a representative sample of the population.

Questionnaire surveys have been used to investigate a range of geographic issues including perceptions of risk from natural hazards, environmental attitudes, gender differences and travel patterns.

To conduct a questionnaire survey, follow the steps below:

1. Define the population to be interviewed e.g. adults, male, female, dependants or chief earner of household.
2. Decide on the sampling method to be used. An appropriate sampling method is required to ensure survey respondents are genuinely representative of the whole population. Three sampling techniques are commonly employed:
   - **a** Random: All members of the defined population have an equal chance of being a respondent to the survey.
   - **b** Systematic: Respondents are selected at regular intervals from a list (e.g. every fourth person on the list).
   - **c** Stratified: Respondents are selected to fit certain criteria (e.g. an even balance of male and female respondents are selected).
3. Decide on a manageable but statistically viable sample size. In most cases, a sample size of more than 30 is necessary to ensure your results are statistically significant.
4. Questionnaires can vary enormously in their design, however to qualify as a valid research questionnaire it should:
   - **a** Collect data in a way that can later be used for analysis.
   - **b** Consist of a predetermined list of questions.
   - **c** Gather information directly from the respondents.
5. Decide on the types of questions to ask. Your questionnaire could include a mix of open and closed questions. Open questions are those that leave the respondent to decide the length and wording of their answer, for example 'What mode of transport do you use to journey to work or school?' Closed questions are those that offer a limited choice of predetermined answers, for example 'How long have you lived at your current address?'
   - **a** Less than one year
   - **b** 1–2 years
   - **c** 3–4 years
   - **d** 5 years or more.
6. Finally, conduct the interview. You can conduct it by telephone or email, however the best approach is face to face.

ISBN: 9780170368155

Guidelines for designing questions

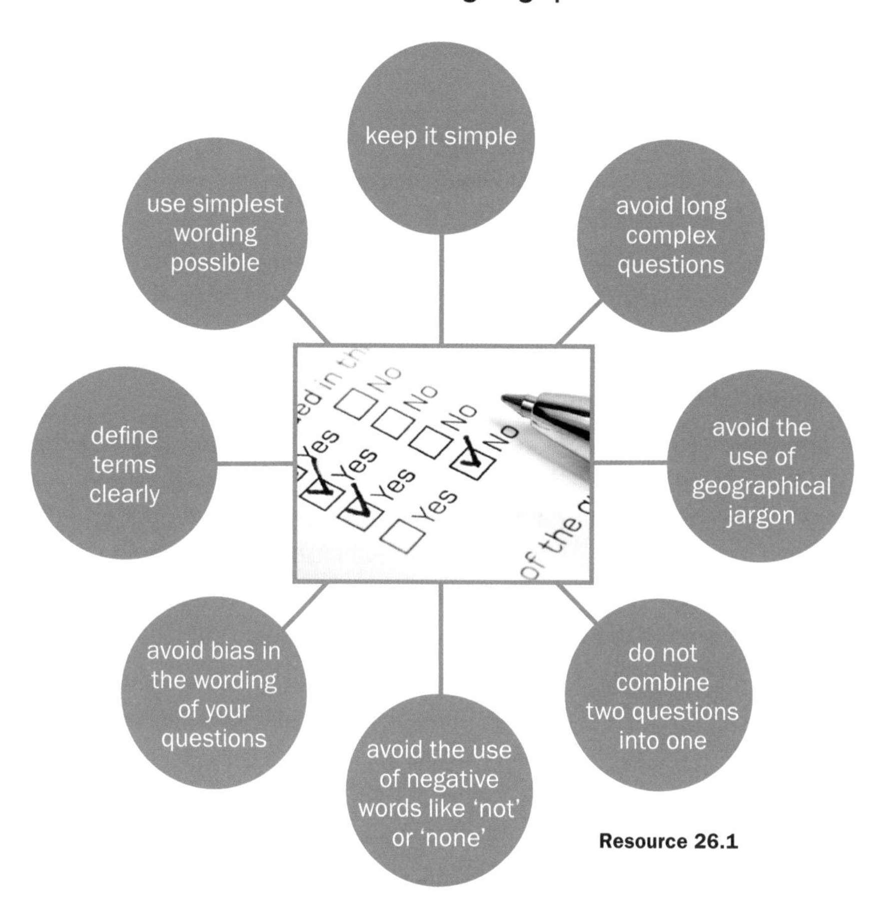

Resource 26.1

Learning Activities

1 Construct a questionnaire to ask tourists visiting popular attractions. Identify where they come from, where they intend to visit in New Zealand and for how long.

a Conduct interviews at a popular tourist destination. Remember to aim for a sample size of at least 30.

b Construct several graphs to illustrate your findings.

c Summarise your findings in a paragraph.

ISBN: 9780170368155

Learning Activities

2 Construct a questionnaire for new immigrants to New Zealand. Your aim is to establish the reasons (i.e. push and pull factors) why they immigrated and their impressions of the country.

a Conduct the interviews with the appropriate people. Remember to aim for a sample size of at least 30.

b Construct several graphs to illustrate your findings.

c Summarise your findings in a paragraph.

ISBN: 9780170368155

27 Field sketches

Geographers use sketches to identify and record features of the landscape. If the sketch is drawn in response to observations made during fieldwork, then it is called a field sketch.

As with précis sketches you do not need artistic ability to draw an effective field sketch, however you do need to demonstrate your understanding of the selected landscape by being able to identify and draw the important features.

When drawn as part of a fieldwork exercise, field sketches can help you to record:

- the natural and built features of a landscape
- urban and rural land use patterns
- transport and communication patterns
- relationships or interactions between features of the landscape.

To construct a field sketch, follow the steps below:

1. Study the scene and choose the area to be included in the sketch.
2. Construct a rectangular frame for your field sketch.
3. Divide the field sketch frame into three equal areas: foreground, middle distance and background (Resource 27.1).

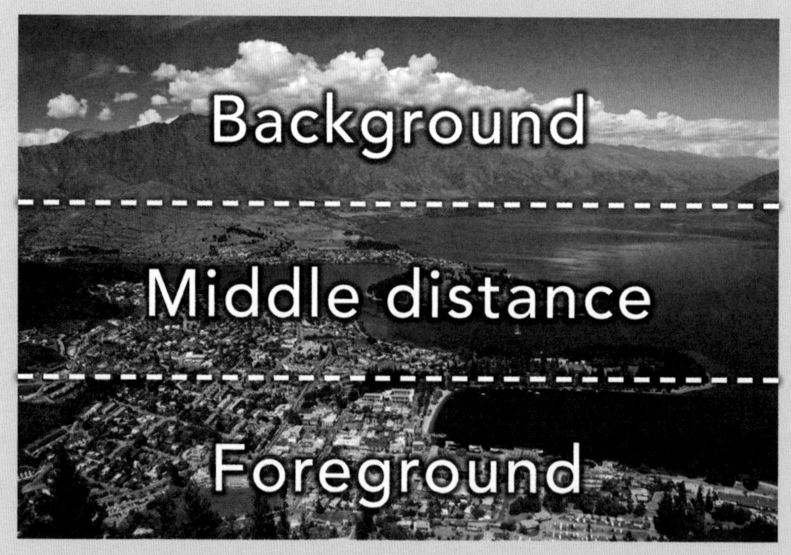

Resource 27.1 Dividing the frame into equal thirds will help you with your sketch

4. Select the main natural and cultural features shown in each area of the scene and sketch the outline of their shape into the frame of the field sketch.
5. Use colour shading to highlight main geographical features.
6. Label the main geographical features.

ISBN: 9780170368155

Resource 27.2 Aoraki/Mt Cook, New Zealand

Resource 27.3 Field sketch of Aoraki/Mt Cook, New Zealand

ISBN: 9780170368155

Learning Activities

1 Explain why geographers construct field sketches as part of their fieldwork.

2 Draw a field sketch below of the view from the summit of Mt Maunganui in Resource 27.4 as though you were sketching the scene in the field.

Resource 27.4

ISBN: 9780170368155

Learning Activities

3 Construct a field sketch of the built environment near your home or school. Label the main geographical features illustrated in your field sketch.

ISBN: 9780170368155

Putting it all together

28 Unit 1: Rotorua relief and recreation

Rotorua is a small inland city set on the southern shores of a lake of the same name. Its full name is *Te Rotorua-nui-a-Kahumatamomoe*; *roto* means lake and *rua* two – its meaning can therefore be translated as 'Second lake'. Recognised for its rich Maori culture, geothermal activity and scenic lakes, Rotorua is a major destination for both domestic and international tourists alike.

Use the maps, graphs and photographs in this section to answer questions 1–3.

ISBN: 9780170368155

Q1 Map and visual image interpretation

Learning Activities

Refer to **Resource A** and **B** to complete the following questions.

Resource A

a Name the type of aerial photograph shown in the resource.

b State two advantages an aerial photograph has over a map.

c State two disadvantages an aerial photograph has compared to a map.

d State the name of the water feature located in the south-eastern quadrant of the photograph.

ISBN: 9780170368155

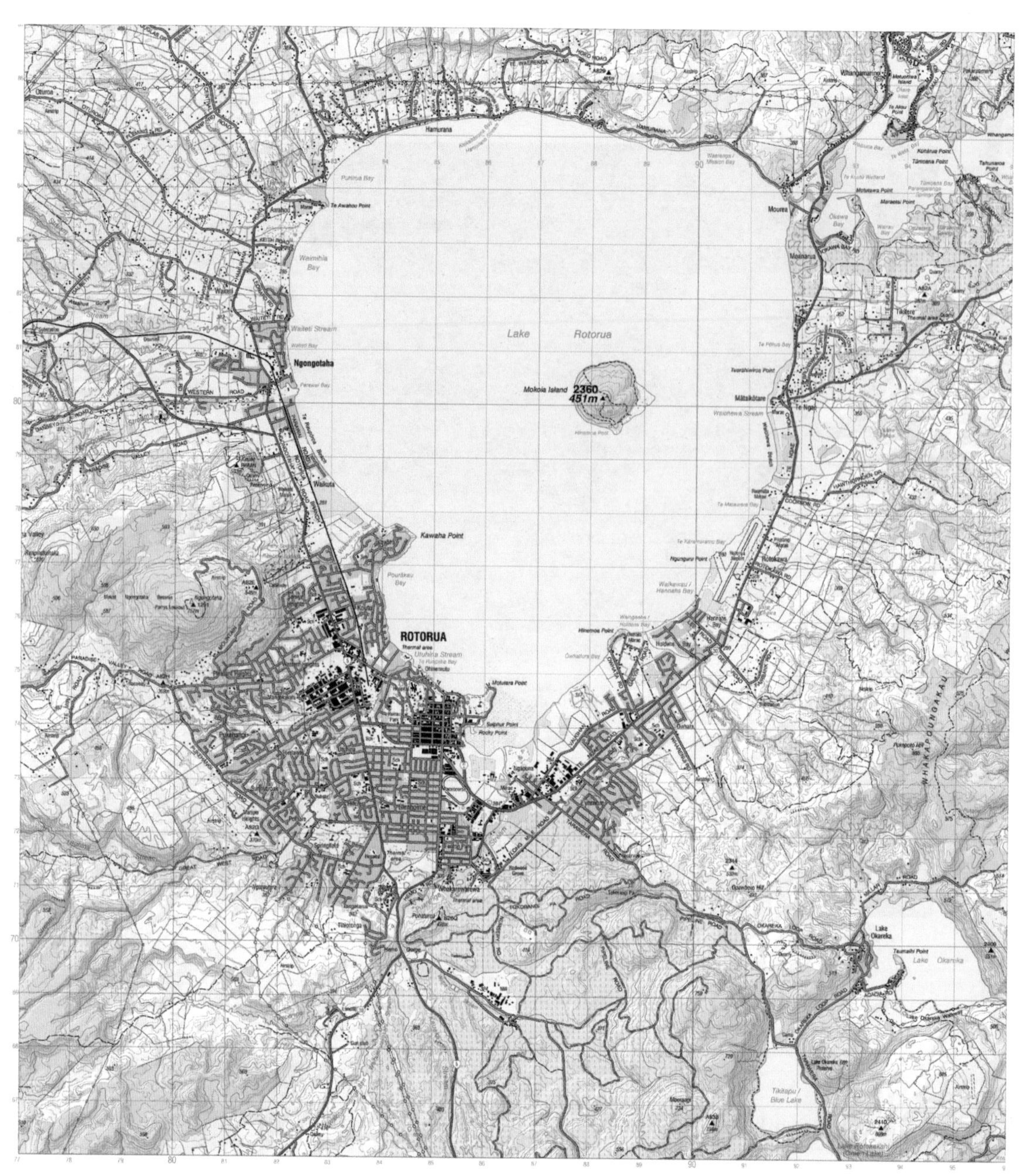

Resource B

Refer to the **Resource B** when answering the following questions.

e On the outline of the précis map of Rotorua opposite locate and label:

- **i** The urban settlement of Rotorua
- **ii** Mokoia Island
- **iii** State Highway 5
- **iv** The transport feature at grid reference 905778.

ISBN: 9780170368155

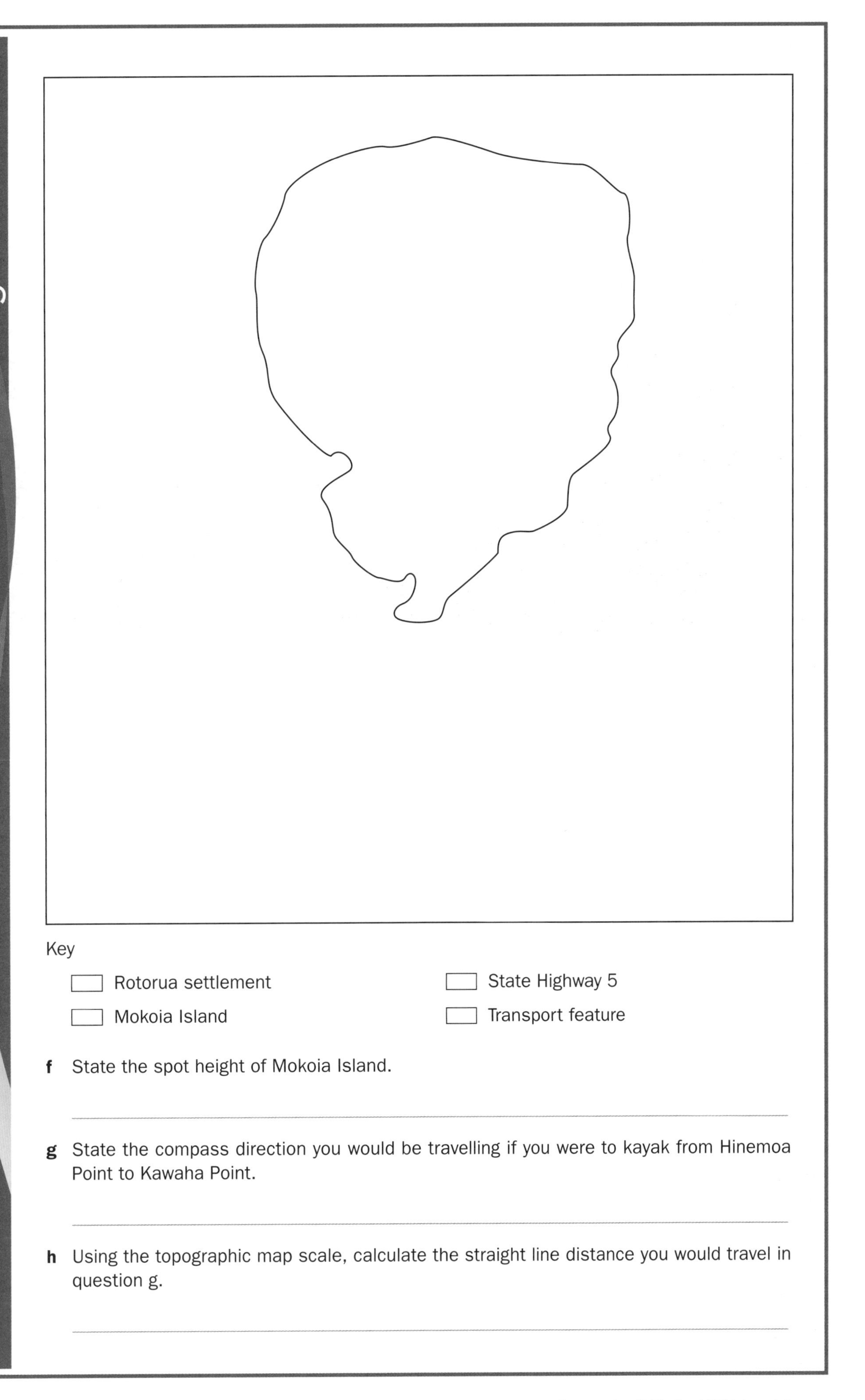

Key

- Rotorua settlement
- Mokoia Island
- State Highway 5
- Transport feature

f State the spot height of Mokoia Island.

____

g State the compass direction you would be travelling if you were to kayak from Hinemoa Point to Kawaha Point.

____

h Using the topographic map scale, calculate the straight line distance you would travel in question g.

____

ISBN: 9780170368155

Refer to **Resource C** and complete the following questions.

Resource C

i Name the type of aerial photograph shown in the resource.

j From what compass direction was the photograph taken?

k State the name of the urban settlement in the foreground.

l Using the frame below, construct a précis sketch of **Resource C** to show:

- **i** Lake Rotorua
- **ii** Mokoia Island
- **iii** State Highway 5
- **iv** The industrial area in the middle distance
- **v** The residential area in the foreground.

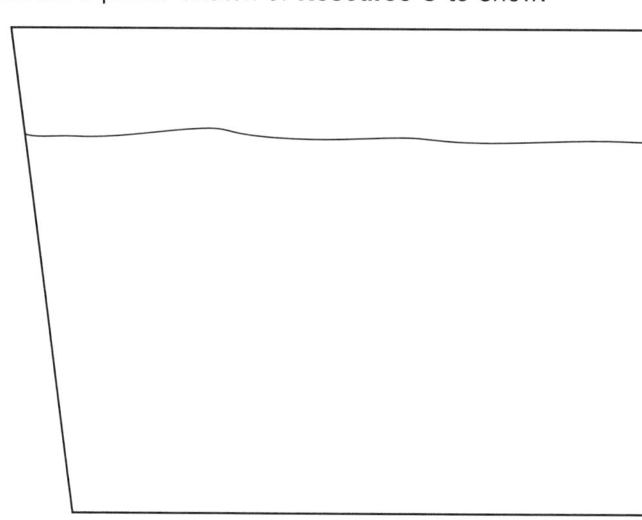

ISBN: 9780170368155

Q2 Graphs and Statistics

Learning Activities

a Refer to **Resource D** and complete the following questions.

Origin of domestic vistors to Rotorua

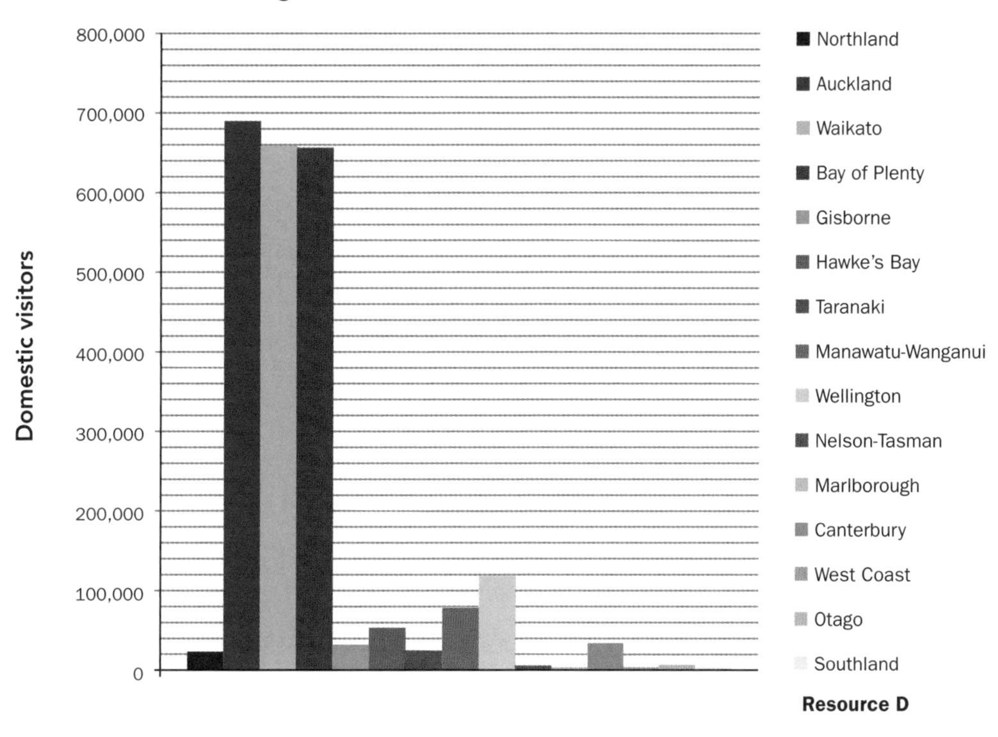

Resource D

i State the largest group of origin for domestic visitors to Rotorua.

ii Suggest why Rotorua receives fewer domestic visitors from the South Island than the North Island.

b Refer to **Resource E** and complete the following questions.

Origin of international visitors to Rotorua (2010)

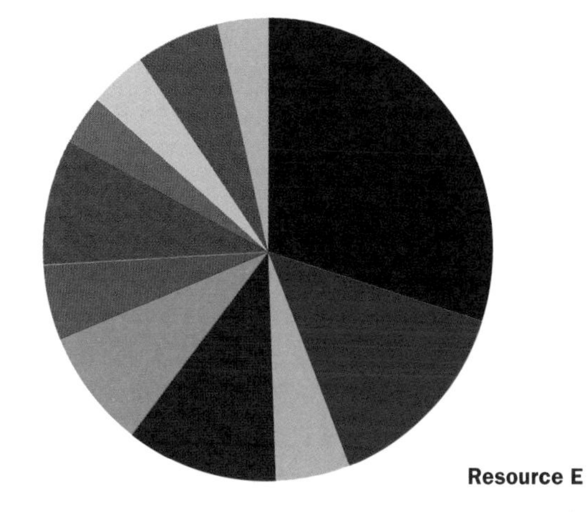

Resource E

- Australia
- United Kingdom
- Germany
- Other Europe
- United States
- Other Americas
- China
- Japan
- South Korea
- Other Asia
- Rest of world

ISBN: 9780170368155

Learning Activities

i Identify the countries of origin of the top four international visitors to Rotorua.

ii With the aid of a protractor, identify the proportion of tourists from the United Kingdom.

iii Suggest why Rotorua receives a large proportion of visitors from Australia?

iv With reference to the resources in the section, write a paragraph about the natural and cultural characteristics of Rotorua's environment that make it an appealing tourist destination.

c Use the data in the table below to construct a climate graph for Rotorua.

	Average rainfall (mm)	Average temperature (°C)
Jan	92.7	17.7
Feb	93.9	17.9
Mar	99.2	16.0
Apr	107.2	13.3
May	116.9	10.7
Jun	136.1	8.5
Jul	134.5	7.8
Aug	131.4	8.4
Sep	109.3	10.2
Oct	112.3	12.0
Nov	93.8	13.9
Dec	114.2	16.2

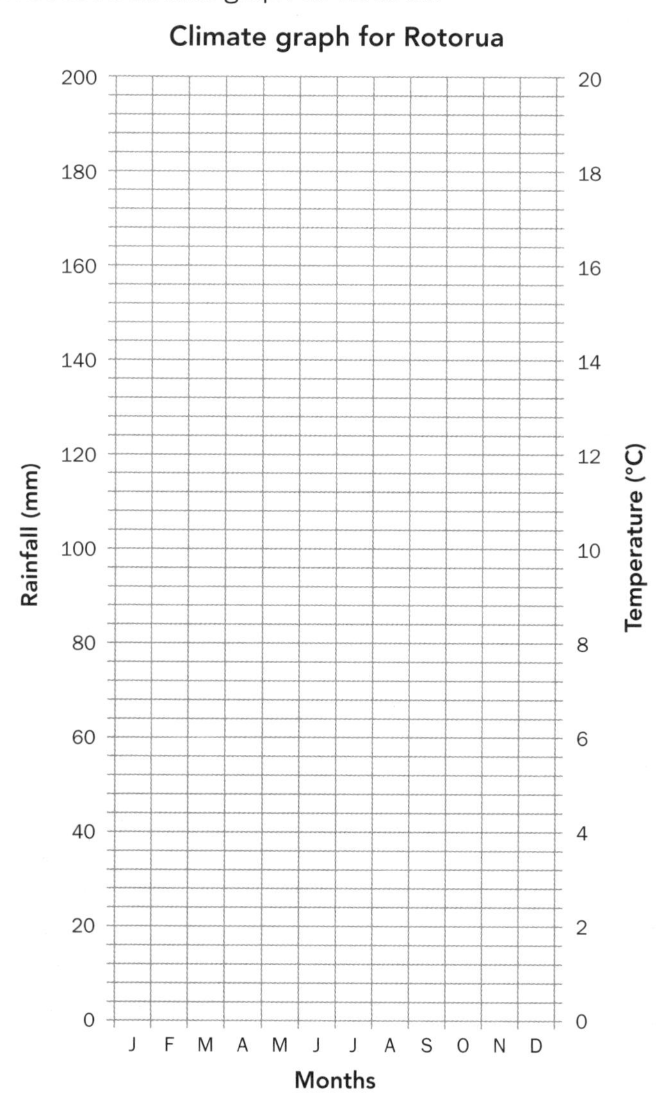

ISBN: 9780170368155

Q3 Application of geographic concepts

Learning Activities

Geographic concepts help us to understand the relationships and connections between people and the natural and cultural environments.

One of the key concepts important to the study of geography is the concept of interaction. Interaction considers the ways the parts (elements) of the environment affect each other or are linked. The natural and built environments are products of interaction.

a With reference to the resources in Unit 1, describe how the geographic concept of interaction is significant to tourism in Rotorua.

b Read the following Maori concept and refer to it, as well as all of the resources in Unit 27 when answering (b).

Maori concept: taonga is a *resource either physical or cultural* that can be found in the environment (including features within the environment e.g. lakes, mountains, rivers, also including people, te reo, whakapapa, etc.).

Explain how the concept of taonga can be applied to tourism in Rotorua. Include specific evidence from the Maori concept and the resources to support your answer.

ISBN: 9780170368155

28 Unit 2: A 'Supercity's' blueprint for growth

At more than 1.5 million people, Auckland is New Zealand's largest and fastest-growing region, accounting for more than half of New Zealand's population growth over the past decade. More than two-thirds of that growth has come from a natural increase in population, while the remaining third has come from net migration.

The phenomenal pace of Auckland's population growth has led to an under-supply of housing to meet demand, and declining affordability and home ownership. To help relieve this, the city council has identified future urban residential areas within the current metropolitan urban limit where fast-track development of affordable housing can take place.

Use the maps, graphs and photographs in this section to answer questions 1–4.

ISBN: 9780170368155

Q1 Graphs and statistics

Learning Activities

Refer to **Resources A** and **B** when answering question 1.

Refer to the table below. For each region of New Zealand, calculate:

a The net change in population for each region between 2006 and 2013.

b The percentage change in population for each region between 2006 and 2013.

Regional Council area	**Census population count**		**Increase or decrease 2006–13**	
	2006	2013	Net change **(a)**	Percent **(b)**
NORTH ISLAND				
Northland Region	148,470	151,689	3219	2.2
Auckland Region	1,304,961	1,415,550		
Waikato Region	380,823	403,638		
Bay of Plenty Region	257,379	267,741		
Gisborne Region	44,496	43,653		
Hawke's Bay Region	147,783	151,179		
Taranaki Region	104,124	109,608		
Manawatu-Wanganui Region	222,423	222,669		
Wellington Region	448,959	471,315		
SOUTH ISLAND				
Tasman Region	44,625	47,157	2532	5.7
Nelson Region	42,888	46,437		
Marlborough Region	42,558	43,416		
West Coast Region	31,326	32,148		
Canterbury Region	521,832	539,433		
Otago Region	193,803	202,470		
Southland Region	90,876	93,339		

Resource A Population by region (Census 2006, 2013)

c Name the three regions that experienced the greatest percentage increase between 2006 and 2013.

____

____

____

ISBN: 9780170368155

d Use the data calculated in question 1b to construct a choropleth map, to display the regional growth of New Zealand's population between 2006 and 2013.

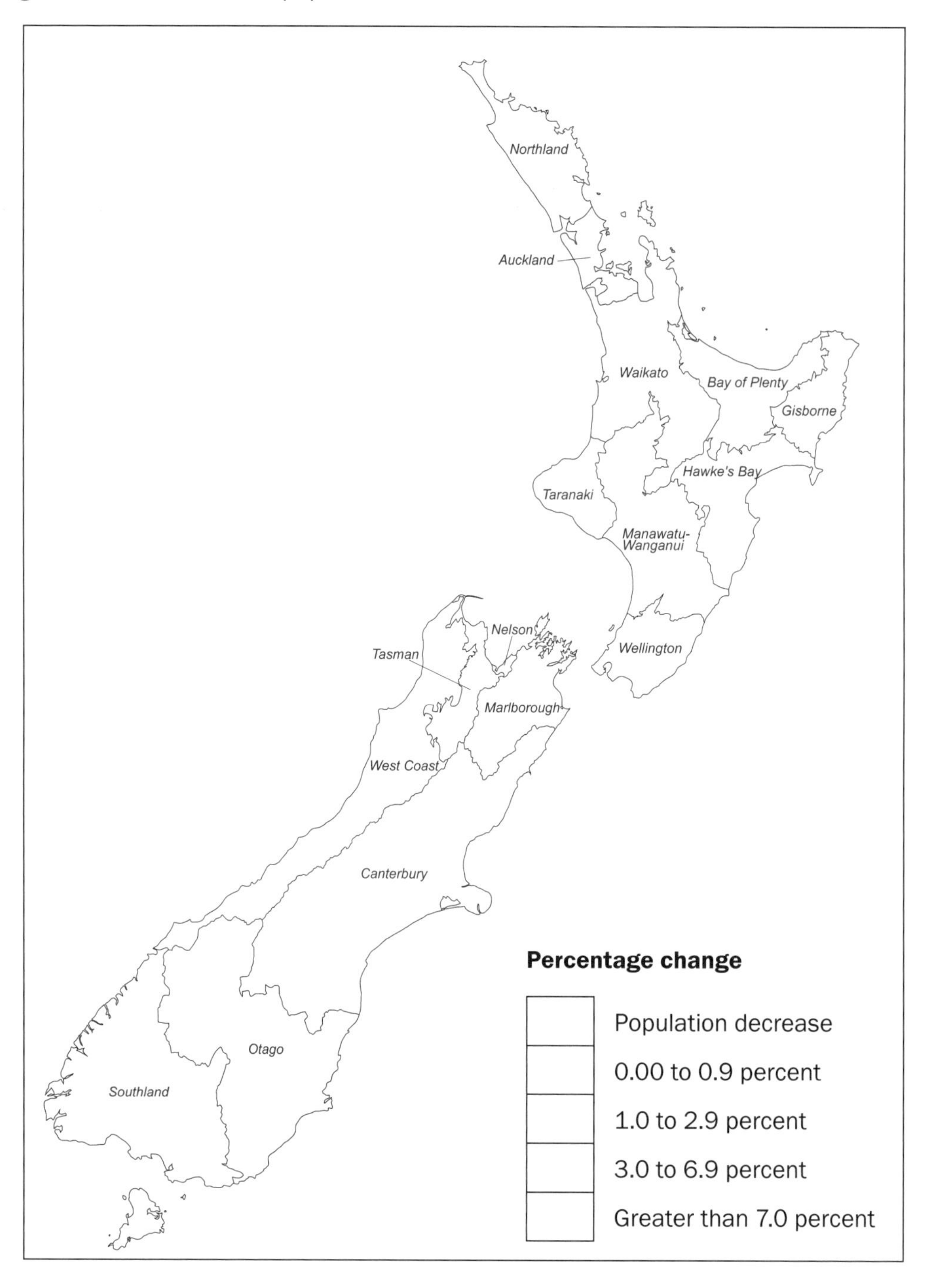

e Describe the pattern of population growth shown in the map.

ISBN: 9780170368155

Learning Activities

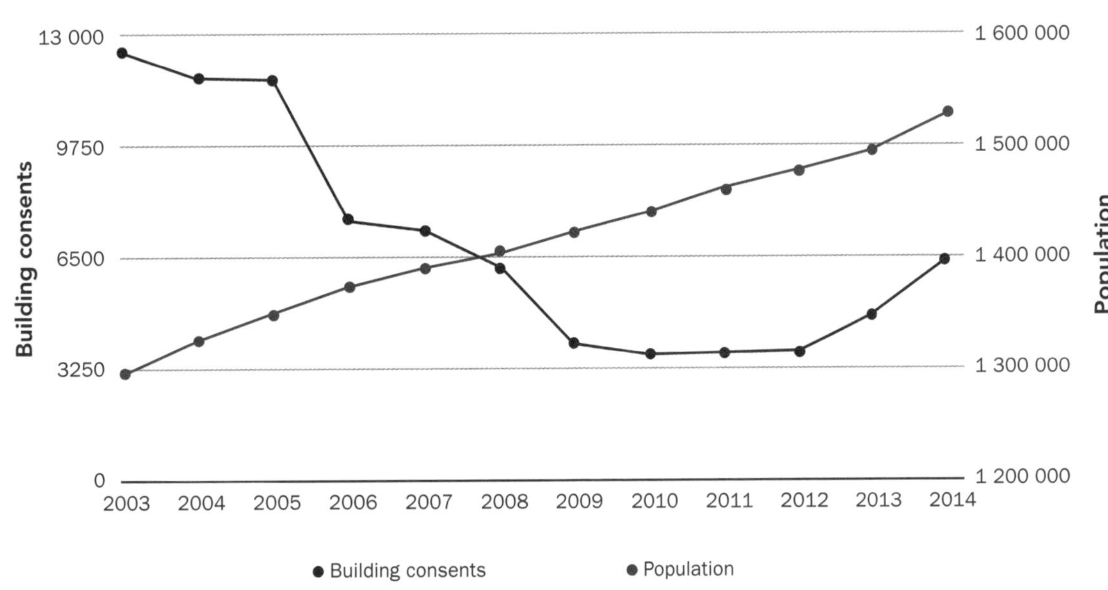

Resource B

f Refer to **Resource B**. Describe changes to Auckland's population growth relative to the growth of building consents issued from 2003 to 2014.

ISBN: 9780170368155

Q2 Visual image interpretation

Learning Activities

Refer to **Resources C – E** when answering question 2.

Resource C 'Mum! Dad! — We've thought of a way you can help us save for a house!'

Context: In June 2013 house prices, especially in Auckland and Christchurch, continued to soar. Savings for a deposit, especially with loan restrictions, seemed to be out of the reach of most families.

a Study the cartoon above.

i Identify the geographic issue (or issues) portrayed in the cartoon 'Mum! Dad! — We've thought of a way you can help us save for a house!'

____

____

____

____

ii What is the perspective of the cartoonist Malcolm Evans?

____

____

____

____

____

ISBN: 9780170368155

Resource D Suburban Auckland

Refer to **Resource D**.

b State the compass direction of the image.

____

____

c Identify the transport network feature in the mid-ground.

____

____

d Name the water feature in the foreground.

____

____

ISBN: 9780170368155

Q3 Map and resource interpretation

Refer to **Resources E** and **F** when answering question 3.

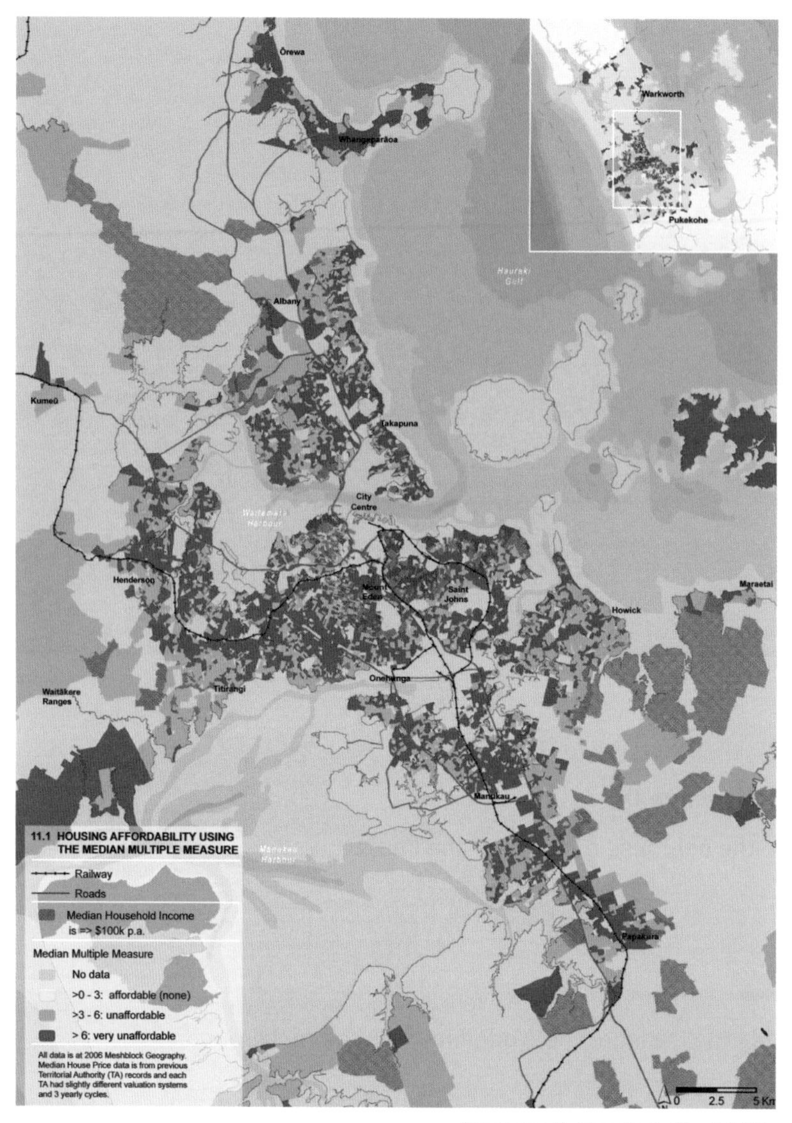

Resource E Housing affordability

Refer to **Resource E**.

a Describe the distribution of areas where the average income is greater than (>) $100,000 per annum.

b Describe the pattern of *very* unaffordable housing.

ISBN: 9780170368155

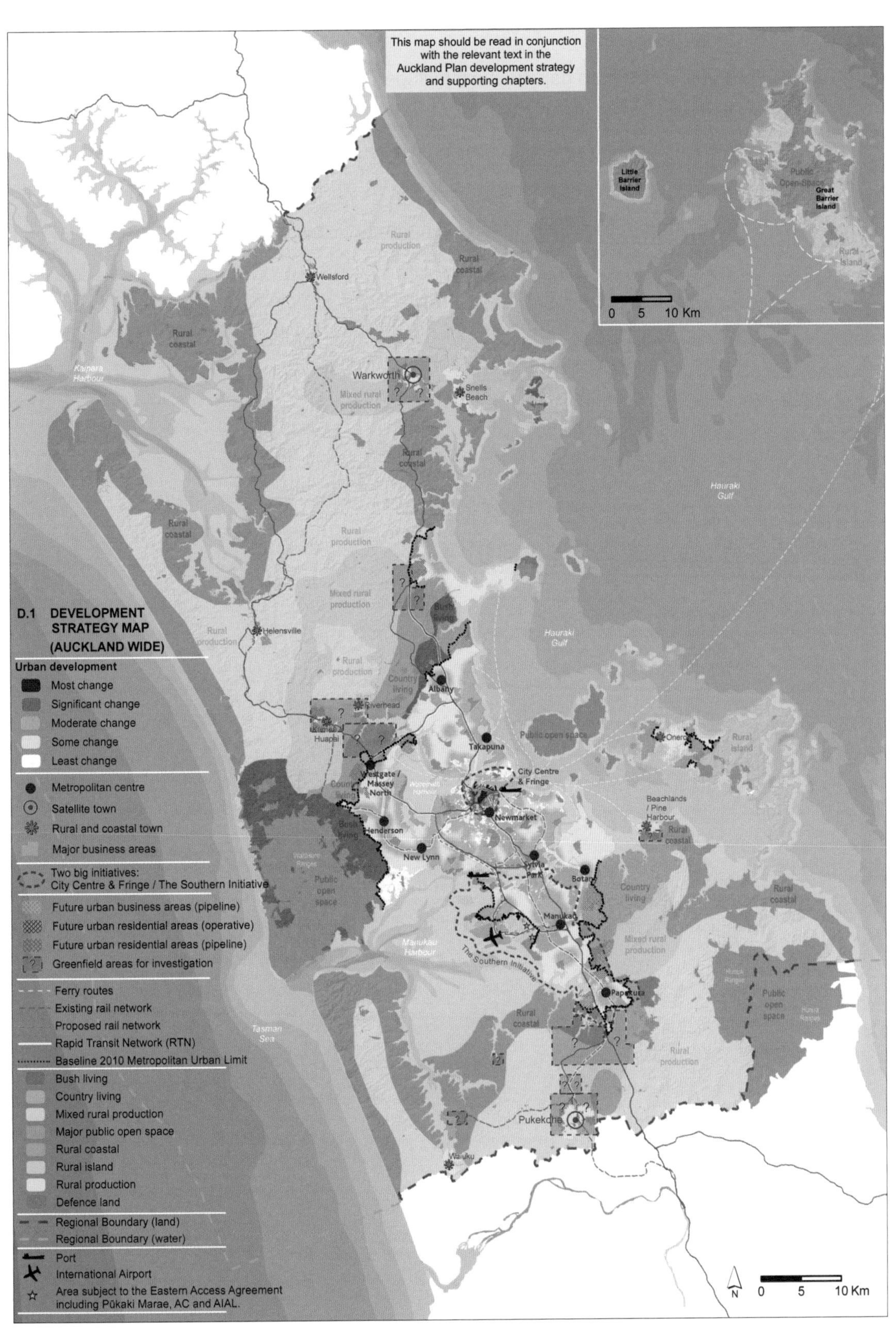

Resource F Development Strategy Map for Auckland

ISBN: 9780170368155

Learning Activities

Refer to **Resource F**.

c Construct a précis map below, which includes the following features:

- **i** The satellite town in the north.
- **ii** The zone of country living to the south-east.
- **iii** The large public open space outside the western metropolitan urban limit.
- **iv** Future urban residential areas within the metropolitan urban limit.

ISBN: 9780170368155

Q4 Application of geographic concepts

Learning Activities

Refer to **Resource G** when answering question 4.

Resource G 'Don't worry about Auckland's heritage, folks.' 18 March 2013

Context: Shows a developer standing on top of a bulldozer, holding up a copy of Auckland's Unitary Plan. He says, 'Don't worry about Auckland's heritage, folks. After all, shoddiness and bad taste are a huge part of our heritage.' Refers to urban planning in Auckland to accommodate the city's ever-growing population.

Geographic concepts help us to understand the relationships and connections between people and the natural environment.

Geographic concept: The concept of sustainability involves ways of thinking and behaving that allow individuals, groups and societies to meet their needs and aspirations without preventing future generations from meeting theirs. Sustainable interaction with the environment may be achieved by preventing, limiting, minimising or correcting environmental damage to water, air and soil, as well as considering ecosystems and problems related to waste, noise, and visual pollution.

a With reference to the cartoon above, fully explain how the geographic concept of sustainability is significant to Auckland's response to its rapid population growth.

Continued on next page

ISBN: 9780170368155

Learning Activities

ISBN: 9780170368155

Learning Activities

Read the following Maori concept and refer to it, as well as all of the resources in Unit 2, when answering question b.

Maori concept: Hekenga refers to *migration* that occurs to meet the needs of Maori at any one time and in response to outside forces.

b Explain how the concept of hekenga can be applied to Auckland's population growth. Include specific evidence from the Maori concept and the resources to support your answer.

ISBN: 9780170368155

Answers

Chapter 1

1 a Teacher to mark this question.

b

	shipwreck
	Maori pa
	cliff
	wind turbine
	underground mine
	index contour
	causeway
	cold spring
	orchard or vineyard

Chapter 2

1 Teacher to mark this question.
2 Teacher to mark this question.
3 a i South-east
ii North-east
iii North-west
b i 315°
ii 225°
iii 135°
4 Teacher to mark this question.
5 a i North-east
ii North-west
iii South-west
iv North-west
b i Approximately 045°
ii Approximately 315°
iii Approximately 225°
iv Approximately 315°
c i Vancouver, Los Angeles, San Francisco
ii Any three destinations from: Auckland, Hamilton, Christchurch, Queenstown, Dunedin, Perth, Adelaide, Melbourne, Sydney, Coolangatta, Brisbane

Chapter 3

1 a i 695025
ii 707013
iii 708999
iv 696017
v 694018
2 a i Racecourse
ii Aran Lodge
iii Church farm
iv Sub station
v Little River Rail Trail
vi Quarry

Chapter 4

1 Teacher to mark this question.
2 Teacher to mark this question.
3

Brisbane, Australia	27°28'S 153°02'E
Dhaka, Bangladesh	23°43'N 90°25'E
Vancouver, Canada	49°16'N 123°07'W
London, England	51°30'N 0°7'W
Berlin, Germany	52°30'N 13°23'E
Reykjavik, Iceland	64°09'N 21°51'W
Kingston, Jamaica	17°58'N 76°48'W
Tokyo, Japan	35°42'N 139°42'E
Auckland, New Zealand	36°50'S 174°44'E
Doha, Qatar	25°15'N 51°36'E

4

Plymouth, UK	50°22'N 4°08'W
Rio de Janeiro, Brazil	22°54'S 43°14'W
Cape Horn, Chile	55°58'S 67°16'W
Tahiti	17°38'S 149°27'W
New Zealand	41°17'S 174°46'E
Botany Bay, Australia	33°58'S 151°10'E
Batavia, Dutch East India	6°12'S 106°48'E
Cape Town, South Africa	33°55'S 18°25'E

5 a 36° 50' S 174° 45' E
b 37° 41' S 176° 9' E
c 39° 5' S 177° 55' E
d 39° 17' S 175° 34'
e 43° 31' S 172° 38' E
f 45° 52' S 170° 30' E
g 45° 1' S 168° 39' E
6 a 4 hours ahead
b 12 hours ahead
c 3 hours ahead
d 7 hours ahead
e 17 hours ahead

Chapter 5

1 a One unit on the map is equal to two million units on the ground
b One unit on the map is equal to 250 000 units on the ground
c One unit on the map is equal to 50 000 units on the ground
d One unit on the map is equal to 2 500 units on the ground
e One unit on the map is equal to 200 units on the ground
2 a 150 000 1:50 000
b 1 250 000 1:250 000
c 11 000 000 1:1 000 000
3 a 2.2 km
b 5.4 km
c 930 m
d 8.7 km

Chapter 6

1 a 160 km^2
b 15 km^2
c 92–98 km^2
d 7 km^2

Chapter 7

1 1 B, 2 E, 3 D, 4 C, 5 F, 6 A
2 Teacher to mark this question.
3 Teacher to mark this question.

Chapter 8

1 Teacher to mark this question.
2 Teacher to mark this question.

Chapter 9

1 a To illustrate the pattern or distribution of a variable or feature.
b The location of cholera deaths.
c Most cholera deaths (red dots) are concentrated around the blue water pump located in the centre of the map (Broad Street).
2 a A map using colour shades to indicate relative values.
b Auckland, Hamilton, Tauranga, Wellington, Christchurch, Dunedin
c Southern Alps, Fiordland, Westland, East Cape, Central North Island, Stewart Island
3 a Isobars show areas of equal air pressure.
b 1032 mb
c 1000 mb
d i Stationary front
ii Warm front
4 a Migration flows
b 1955-1960: New York (NY), Michigan (MI), Texas (TX).
1995-2000: California (CA)

ISBN: 9780170368155

c East Coast to West Coast
d West Coast to East Coast

5 a London
b The population of the south-west is relatively larger than that of the north.

6 a To communicate the variety of ski trails at Coronet Peak.
b It does not have a scale, north point or show a bird's-eye view.
c i Advanced
ii Beginner
iii Intermediate
d Intermediate, advanced

Chapter 10

1 a 1.34 billion
b 881 million

2 a 4000
b Auckland
c West Coast, Gisborne
d Natural increase

3

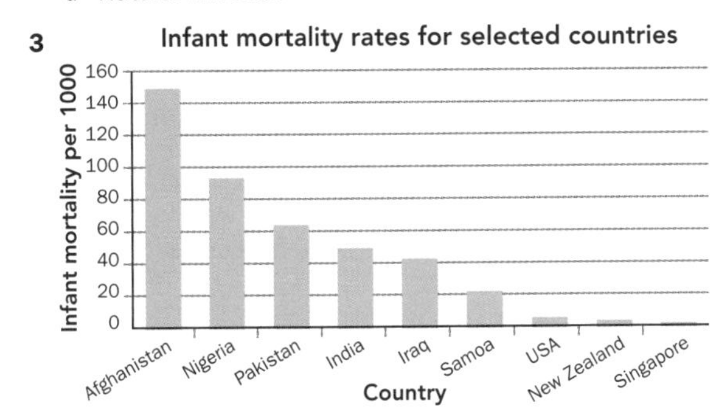

4

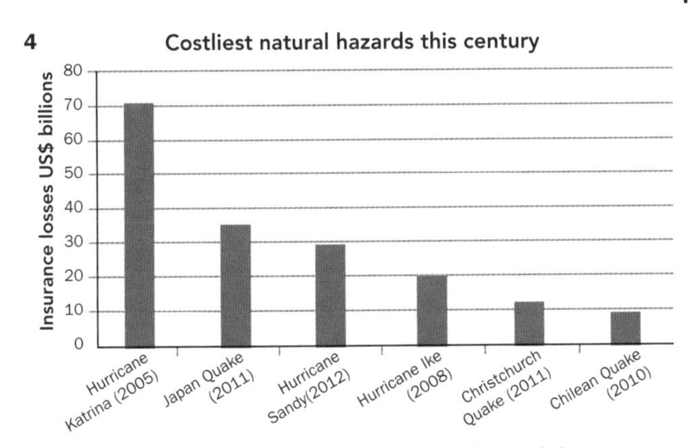

5

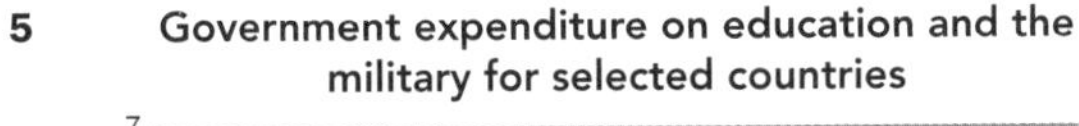

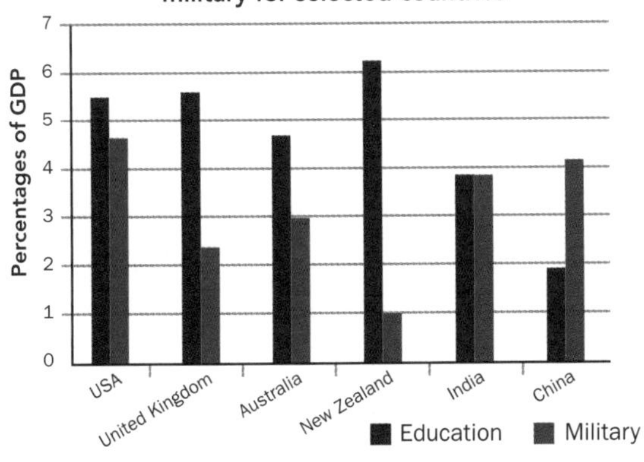

Chapter 11

1 a i 1 billion
ii 1.8 billion
iii 5.4 billion
b 9.4 billion

2 a i 38 per 1000
ii 33 per 1000
iii 15 per 1000
b i 31 per 1000
ii 9 per 1000
iii 7 per 1000
c 1911: 5; 1961: 24; 2011: 8
d i 1953
ii 1911
e Sri Lanka's natural population increase was low in 1911 as the number of deaths nearly matched the number of births. The natural increase rose significantly during the 20th century only to decline again in the 1990s and 2000s.

3

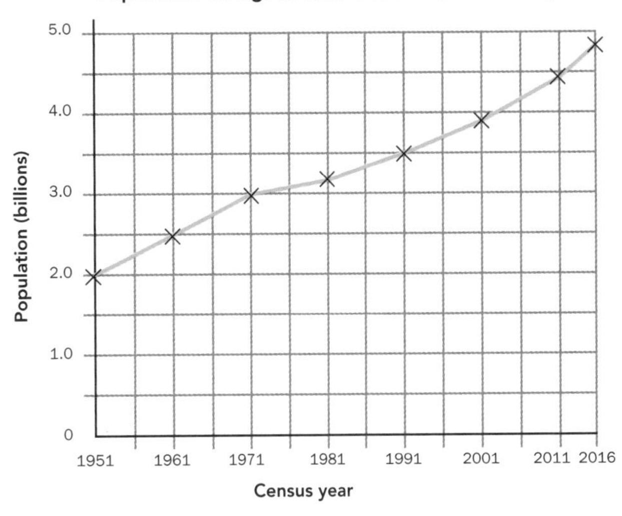

4

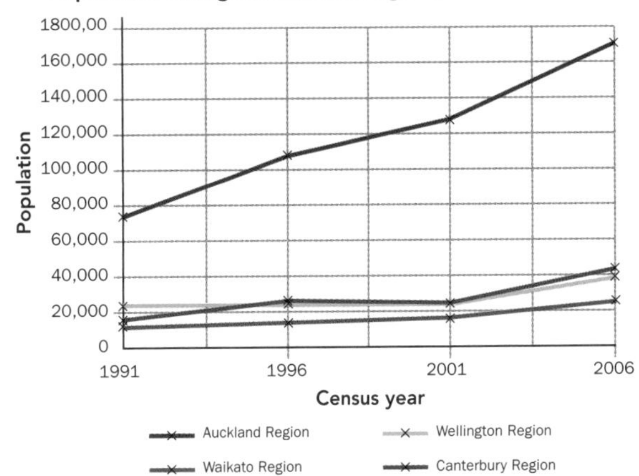

Chapter 12

1 a Hydro
b i Coal, gas
ii Hydro, geothermal, wind
c i 28%
ii 68%

2 Ethnic population of New Zealand (2013)

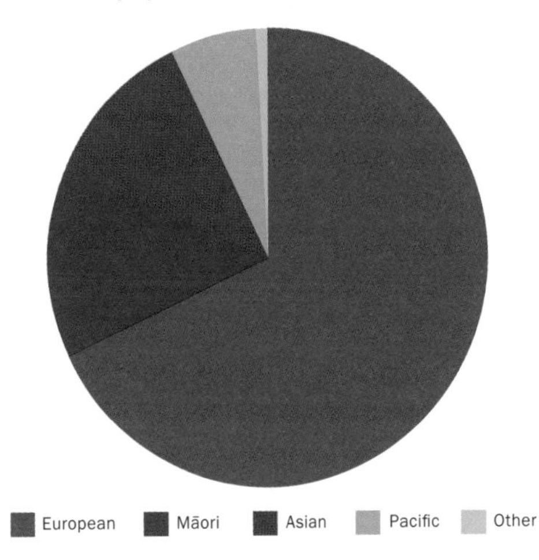

ISBN: 9780170368155

3

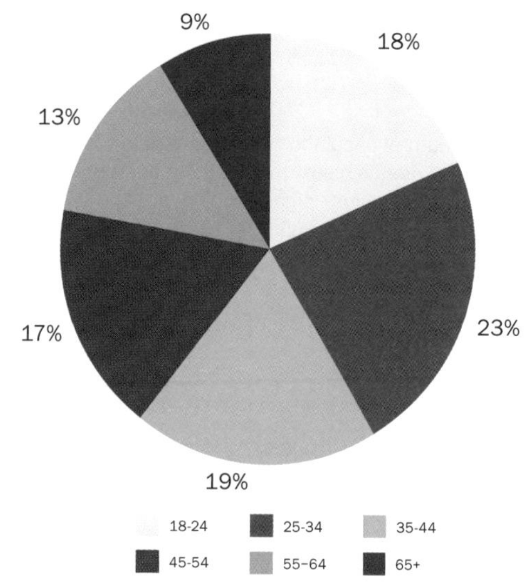

4 a i 60%
ii 14%
iii 8%
iv 0.5%
b i 29.5%
ii 20.4%
iii 16.5%
iv 5.9%
c The population density of Asia is higher than that of Australia.

5

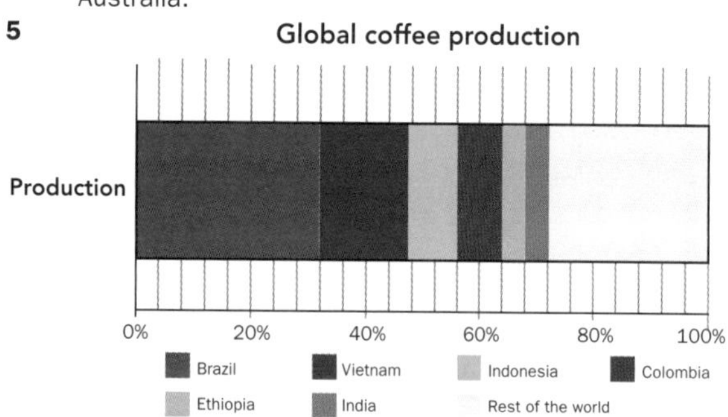

6

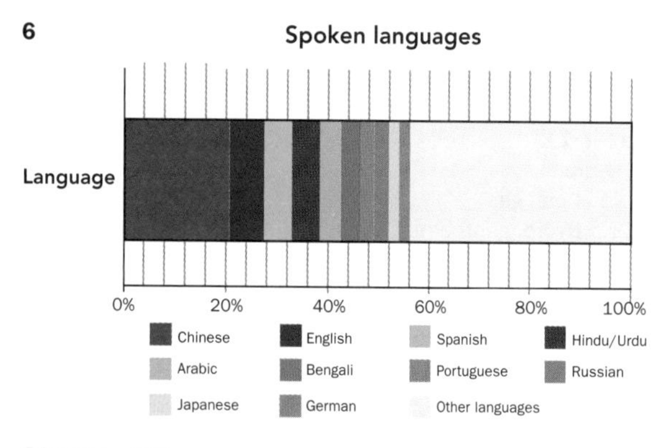

Chapter 13

1 a i Country A 44 years US$305
ii Country B 64 years US$1250
iii Country C 72 years US$8000
b As average income per person increases, so too does the average life expectancy of the population i.e. the correlation between the two variables is positive.

2 a b

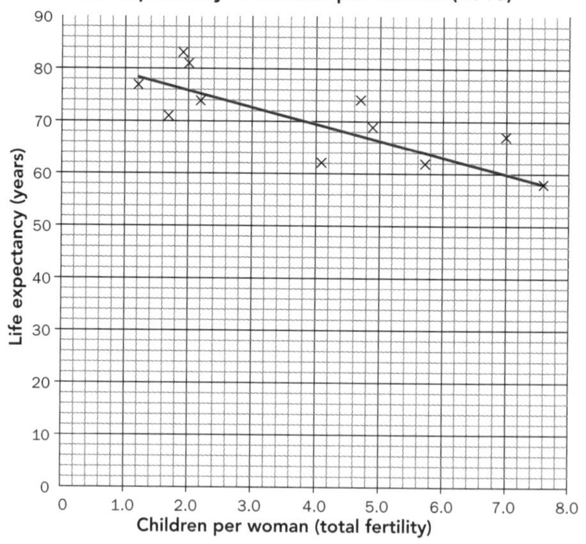

c As the number of children per woman increases, the life expectancy of the population decreases i.e. the correlation between the two variables is negative.

3

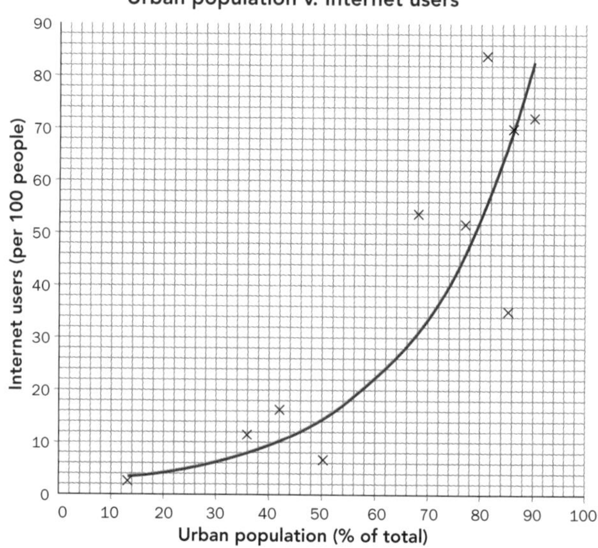

c As the urban percentage of the population increases, so too does the number of Internet users per 100.

Chapter 14

1 a i 4.3
ii 2.5
iii 1
iv 0.3

2

ISBN: 9780170368155

3

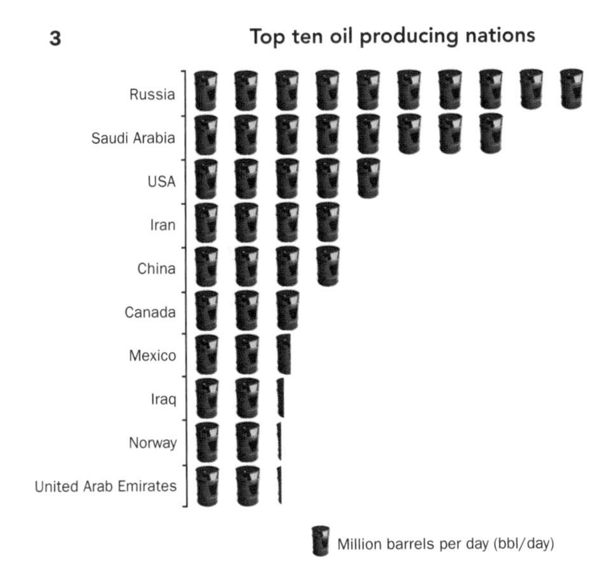

4

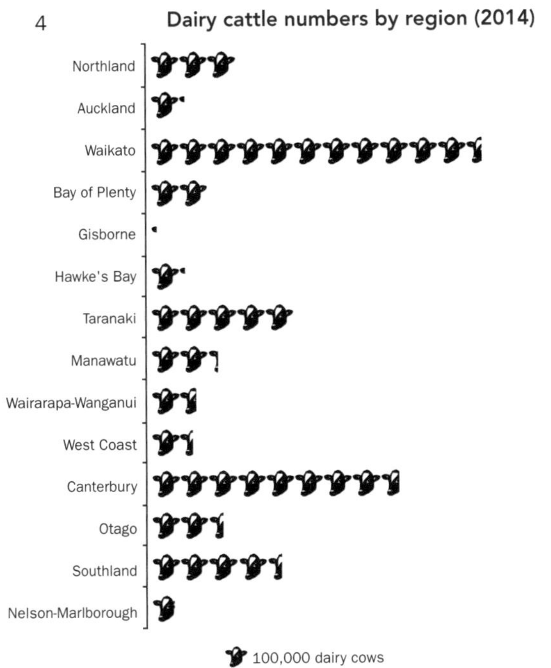

b The majority of dairy cows are located in the Waikato Region followed by the Canterbury and Taranaki Regions.

Chapter 15

1 a i 18.2˚C 80 mm
ii 14.5˚C 95 mm
iii 8.8˚C 126 mm
iv 13.1˚C 96 mm

b Wettest: July. Driest: February

c Coolest: July. Warmest: February

d Hamilton experiences warmer weather during the summer months of December to March. It receives moderate rainfall all year round with the wettest months being from March to October.

2

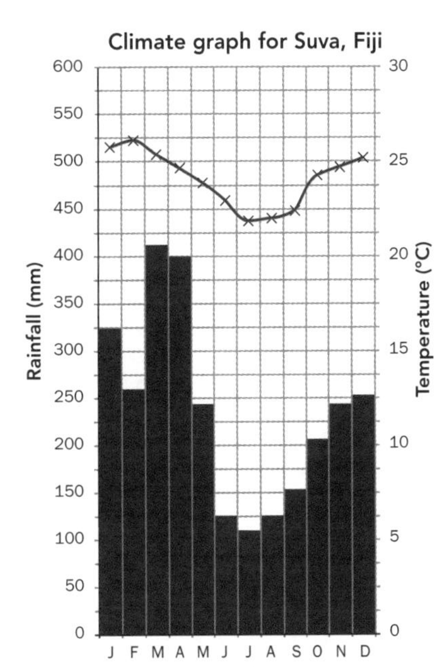

3 Fiji experiences a hot-wet season from November through to April. During this period, both rainfall and temperature are higher than the rest of the year. Conversely, Fiji experiences a cool-dry season from June to September when both rainfall and temperatures are considerably lower.

Chapter 16

1 a 13%

b 23%

c Contracting

2

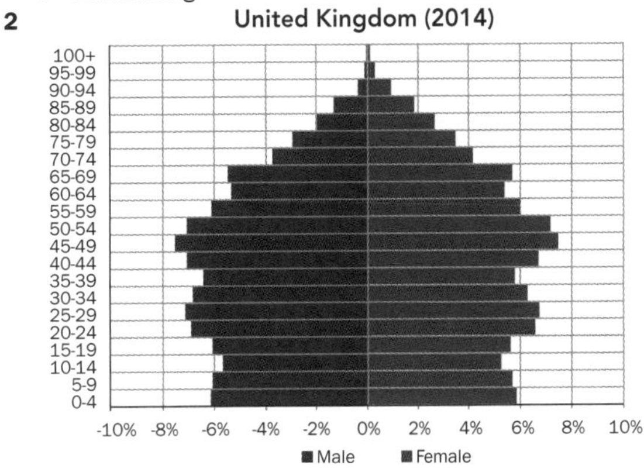

3 The population pyramid for the United Kingdom has a relatively narrow base and a large proportion of people in the 65+ age groups.

4

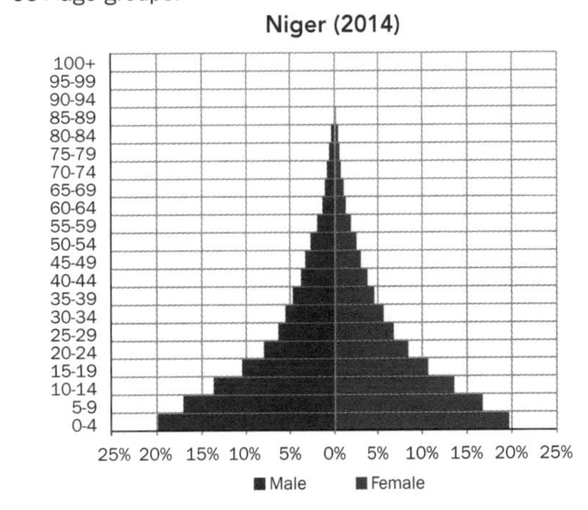

5 The population pyramid for Niger has a wide base and a tapered (narrow) top.

ISBN: 9780170368155

Chapter 17

1 a Resource 17.4
  b Resource 17.5
  c Resource 17.3

2 Features are easier to identify in oblique aerial photographs as objects can be seen from the side. Unlike vertical aerial photographs, oblique aerial photographs do not have a constant scale, making measurement of distances difficult.

3 a i Flat land is found near the coastline. Mountains (volcanic extrusions) are also located near the coast.
    ii Vegetation is found in the hill country and mountainous areas.
    iii The landscape is coastal as demonstrated by the presence of a bay or harbour.
    iv The settlement is located on the lowlands adjacent to the sea.
  b Teacher to mark this question.

4 Teacher to mark this question.

Chapter 18

1 To identify and record features of a landscape.
2 Teacher to mark this question.
3 Teacher to mark this question.

Chapter 19

1

contemporary issue	An issue that is current; contentious or open to debate
symbolism	The use of symbols to represent ideas or qualities
caricature	A picture, description, or imitation of a person or thing in which certain characteristics are exaggerated to create a comic or grotesque effect
stereotype	A popular belief about specific social groups or types of individuals
visual metaphor	The representation of a person, place, thing, or idea by way of a visual image that suggests a particular association or point of similarity
perspective	The way people view and interpret the world around them

2 Teacher to mark this question.
3 Teacher to mark this question.
4 a The 19th century European settlers to New Zealand
  b Little consideration was given to potential hazard risks
  c i Auckland
    ii Christchurch, Wellington, Napier, Hastings
    iii Christchurch, Wanganui

Chapter 20

1 Land use zoning is a strategy used by local government to control or restrict the use of land from certain purposes or activities. With this strategy in mind, Figure 20.1 shows how flood plains, which have the potential to flood, should only be used for parks and reserves, while larger structures such as residences and high rise buildings should be built well above potential flood levels. Such a strategy, if observed, will ensure that any future flood events will have a minimal impact on people and property.

2 Teacher to mark this question.

3

The Location of North Island Volcanic Fields

Northland (dormant) - The most recent eruption was Mt Te Puke (near Waitangi) about 1,500 years ago.

Auckland (dormant) - There are about 50 volcanoes or craters in the central Auckland area. The last eruption was Rangitoto Island about 750 years ago. An eruption in the area today would be harmful for Auckland's large, expansive population.

White Island (active) - A major eruption could potentially cause a tsunami threat to the Bay of Plenty coast.

Taranaki (dormant) - There have been at least nine eruptions in the last 500 years. The main hazard is ash.

Okataina (dormant) - In the last 250,000 years there have been around five major eruptions. One of the most devastating was Tarawera in 1886.

Tongariro (active) - This area includes Ngauruhoe, Ruapehu and Tongaririo. Ruapehu erupted as recently as 2007, producing lahars. Ngauruhoe last erupted in 1975.

Taupo (dormant) - The lake is a giant caldera caused by a series of huge eruptions ending about 1850 years ago when everything around the vent for 20,000 sq km was destroyed.

Chapter 21

1 Teacher to mark this question.

Chapter 22

1 a i 7
    ii 2
    iii 4
    iv 1
  b North-west
  c The wind at Palmer Station, Antarctica generally blows from the north (7 days) to north-west (8 days).

Chapter 23

1 Stage 1 = E
  Stage 2 = D
  Stage 3 = B
  Stage 4 = A
  Stage 5 = C

2 **Population change in Sri Lanka**

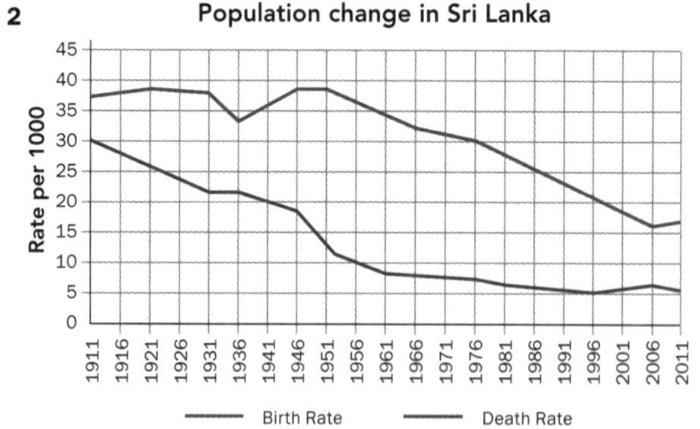

3 a 1911 - 1931
  b 1931 - 1961
  c 1961 - 2011

4 The evidence in the graph suggests that Sri Lanka is now in Stage 3 of the DT model. Stage 3 of the model is characterised by low death rates, falling birth rates and high population growth.

ISBN: 9780170368155

Chapter 24

1

Push factors	Pull factors
Not enough jobs	Stable governance
Loss of wealth	Quality education
Slavery or forced labour	Better medical care
Political fear or persecution	Attractive climates
Primitive living conditions	Security
Desertification	Job opportunities
Natural disasters	Political or religious freedom
Pollution	Better chances of marrying
Unsafe housing	Family links
Lack of political or religious freedom	
Civil war	

Chapter 25

1 Teacher to mark this question.
2 Teacher to mark this question.
3 Teacher to mark this question.
4 Teacher to mark this question.

Chapter 26

1 Teacher to mark this question.
2 Teacher to mark this question.

Chapter 27

1 To identify and record features of a landscape.
2 Teacher to mark this question.
3 Teacher to mark this question.

Chapter 28 – Unit 1

1 a Vertical aerial photograph.
b Objects are easy to identify, not over generalised or simplified, spatial patterns are more evident as are the relationships between natural and cultural features.
c The map gives an indication of elevation, includes the names of places and features, easier to calculate distance.
d Lake Tarawera
e Teacher to mark this question.
f 451 m
g North-west
h 4.25 km
i Oblique aerial photograph.
j South-west
k Rotorua
l Teacher to mark this question.

2 a i Auckland
ii Due to the cost and time it takes to travel (i.e. distance decay).
b i Australia, United Kingdom, Germany, United States
ii 14%
iii Close proximity to New Zealand; favourable exchange rate.
iv Teacher to mark this question.

c

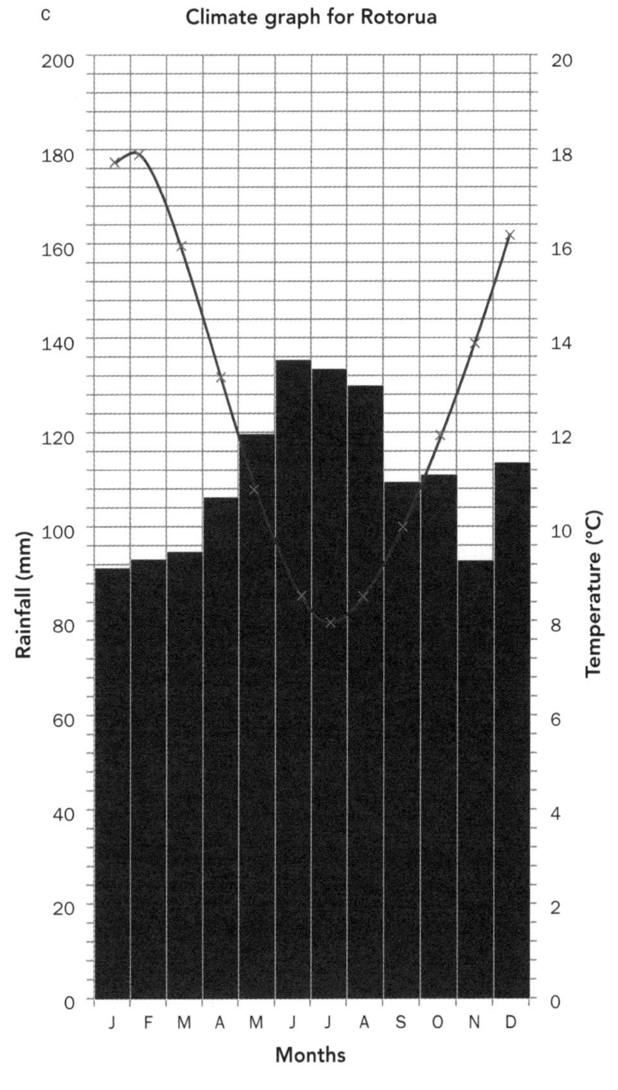

3 a Teacher to mark this question.
b Teacher to mark this question.

Chapter 28 – Unit 2

1 a

Regional Council Area	Net Change
North Island	
Northland Region	3219
Auckland Region	110589
Waikato Region	22815
Bay of Plenty Region	10362
Gisborne Region	-843
Hawke's Bay Region	3396
Taranaki Region	5484
Manawatu-Wanganui Region	246
Wellington Region	22356
South Island	
Tasman Region	2532
Nelson Region	3549
Marlborough Region	858
West Coast Region	822
Canterbury Region	17601
Otago Region	8667
Southland Region	2463

ISBN: 9780170368155

b

Regional Council Area	Percent
North Island	
Northland Region	2.2
Auckland Region	8.5
Waikato Region	6.0
Bay of Plenty Region	4.0
Gisborne Region	-1.9
Hawke's Bay Region	2.3
Taranaki Region	5.3
Manawatu-Wanganui Region	0.1
Wellington Region	5.0
South Island	
Tasman Region	5.7
Nelson Region	8.3
Marlborough Region	2.0
West Coast Region	2.6
Canterbury Region	3.4
Otago Region	4.5
Southland Region	2.7

c Auckland, Nelson and Waikato

d

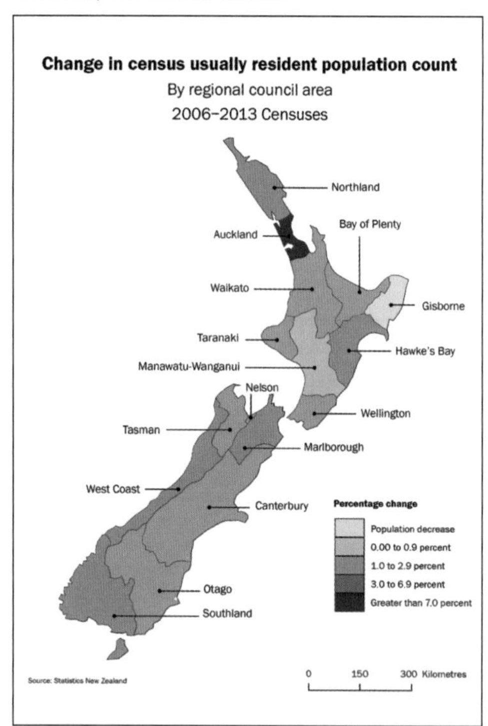

e The greatest growth between the 2006 and 2013 census occurred in the regions that contain major urban areas (e.g. Auckland, Hamilton, Tauranga, Wellington, Christchurch and Dunedin). Manuwatu-Wanganui experienced the least growth while the Gisborne region decreased in population.

f The building of new houses as measured by building consents issued has failed to keep pace with Auckland's population growth. The likely result is a shortage in housing and an increase in house prices.

2 a i An increase in population and decrease in housing supply has decreased the affordability of housing in some areas. As a result, saving for a deposit and consequently buying a home is out of reach for many families and often the only option families have is to continue to live with their parents or other family members.

ii Possible answer: Empathy towards the parents or other family members who carry the burden of hosting relatives, long term; empathy for young families who have been priced out of the housing market.

b East

c Bridge (Harbour Bridge)

d Waitamata Harbour

3 a The highest incomes are mainly found in the City centre and in the rural areas to the north and south-east of urban Auckland.

b Scattered throughout urban Auckland.

c Teacher to mark this question.

4 a Teacher to mark this question.

b Teacher to mark this question.

ISBN: 9780170368155